CREATIVELY UNDECIDED

CREATIVELY UNDECIDED

Toward a History and Philosophy of
Scientific Agency

MENACHEM FISCH

THE UNIVERSITY OF CHICAGO PRESS
CHICAGO AND LONDON

The University of Chicago Press, Chicago 60637
The University of Chicago Press, Ltd., London

Published 2017
Printed in the United States of America

26 25 24 23 22 21 20 19 18 17   1 2 3 4 5

ISBN-13: 978-0-226-51448-2 (cloth)
ISBN-13: 978-0-226-51451-2 (paper)
ISBN-13: 978-0-226-51465-9 (e-book)
DOI: 10.7208/chicago/9780226514659.001.0001

Library of Congress Cataloging-in-Publication Data

Names: Fisch, Menachem, author.
Title: Creatively undecided : toward a history and philosophy of scientific agency / Menachem Fisch.
Description: Chicago : The University of Chicago Press, 2017. | Includes bibliographical references and index.
Identifiers: LCCN 2017014001 | ISBN 9780226514482 (cloth) | ISBN 9780226514512 (pbk. : alk. paper) | ISBN 9780226514659 (e-book)
Subjects: LCSH: Science—Philosophy. | Ambiguity in science. | Mathematics—England—History—19th century. | Science—History—19th century.
Classification: LCC Q175.F5452 2017 | DDC 501—dc23
LC record available at https://lccn.loc.gov/2017014001

∞ This paper meets the requirements of ANSI/NISO Z39.48-1992 (Permanence of Paper).

Dedicated with love and gratitude to the Cohn Institute for the History and Philosophy of Science and Ideas, Tel Aviv University; its benefactors, Bert and Barbara Cohn; its founders, the late Yehuda Elkana and Amos Funkenstein; and the colleagues, staff, and students who have made it my philosophical home.

CONTENTS

INTRODUCTION

Our understanding of paradigm shifts—or, as I prefer to term them following Friedman, scientific framework transitions—has been caught since the late 1950s between two impossibly limiting positions. On the one hand, there is the critical rationalism of Karl Popper and his school, which fails to properly distinguish between first-order scientific work and the set of normative commitments within which it is inevitably couched; on the other, the account of scientific framework transitions proposed by Thomas Kuhn and his school, which fails to present them as capable of being undertaken rationally. The present study is an attempt to follow the latter's neo-Kantianism in arguing for a strong notion of framework dependency of all scientific practice and product, and at the same time to show how such frameworks can be deemed the possible outcomes of keen rational deliberation, much along Popperian lines, again in the strongest sense of the term—a combination of philosophical commitments that has so far been regarded as oxymoronic.

Michael Friedman's *Dynamics of Reason* (2001) stands out in being the only philosophical attempt of note to address the problem of the rationality of such transitions squarely from within the neo-Kantian school.[1] The account of framework transitions expounded and argued for in the first four chapters of the present book, and put to the test of a detailed case study in the remaining three, takes issue with major aspects of Friedman's account. For one thing, it defines the problem of the rationality of such transitions very differently. Friedman sees the problem as one of communal rational *choice*, a matter of explaining how practitioners can rationally deem a constitutive framework radically different from their own to be a "live option" from their perspective—a question that can arise only *after* such a framework is in place and has won supporters. The problem with which the present study contends is not primarily that of communicating rationally across an already existing divide,[2] but that of articulating the rational *incentive* to create an alternative framework in the first place. The question is how to account *prospectively* for the reasons practitioners might have for initially attempting to *devise* a framework radically different from the one to which they are committed.

Even if a person could be rationally won over by a framework she deems superior to her own (as Friedman believes one can, and as the present study denies),[3] to endeavor rationally to seek or develop such an alternative in the first place is a different matter entirely. To do so, practitioners must be able to deem the normative framework to which they are committed to be normatively wanting. But that would seem impossible. How can a person subject her scientific standards to rational appraisal if it is by means of those very standards that she rationally appraises? How can a constitutive framework be indicted by the very principles of rational reckoning to which it gives rise? How could a committed Aristotelian justify replacing his Aristotelian framework by force exclusively of Aristotelian argument?

Normative criticism necessarily requires prior commitment to a framework of norms and standards that, by virtue of that very commitment, is exempted from the type of normative criticism it makes possible. It is, therefore, seemingly impervious in principle to rational appraisal from within. Certain of our normative commitments weaken and fade, and are replaced unthinkingly—as is usually the case with our aesthetic tastes and standards. But we would like to think that when we change the normative basis for our ethical, political, religious, and scientific judgments we can, at least in principle, be said to do so rationally, that is, for a reason. To claim that we cannot strikes us as philosophically preposterous. And yet, as I shall argue in chapter 2, it is a problem that has stumped the contemporary neo-Kantian discussion of normativity, self, and agency in general, and is responsible for the vexing impasse in the study, philosophical and historical, of scientific framework transitions, as argued further in chapter 3.

The reason the problem strikes us as preposterous is that, although we realize the impossibility of pronouncing our very standards of wrongness to be wrong by merely talking to ourselves, we all know from personal experience that exposure to the normative critique of others can have that transformative effect. Accounting for it, as we shall see, is not easy. But it is not as if philosophers have tried unsuccessfully to do so and have failed. They have simply not tried! What has stumped the philosophical discussion in this regard is a similar and related, if less contentious, divide that exists in contemporary philosophy between such thinkers as Christine Korsgaard,

Harry Frankfurt, and Charles Taylor, who offer rich accounts of agency and self and of the intrasubjective realm of normative self-deliberation,[4] and such thinkers as the later Wittgenstein, Robert Brandom, and Michael Walzer, whose work (much like Friedman's) focuses exclusively on the intersubjective realm of normative discourse.[5] As a result, the interface between, and the mutual bearings of, what might be termed (following Wilfrid Sellars, John McDowell, and Brandom) the "public space of reasons," where the intersubjective game of giving and asking for reasons is played, and the private, intrasubjective space of reasoned pondering is simply nowhere explored.

Brandom's recent work is a case in point, as it would seem the exception. His Woodbridge Lectures of 2008 present the most developed version of his account of rationality as a "Semantic Sonata" in which a Kantian account of intrasubjective normative self-binding is supplemented as a matter of necessity by a Hegelian account of intersubjective "reciprocal recognition."[6] This is the only work I know on self and agency in which the discursive realms of the personal and the social are truly and systematically juxtaposed and joined. Brandom's Kantian chapter focuses on the intrasubjective management of one's commitments, on the way we constitute ourselves as subjects by sustaining and developing a synthetic unity of apperception, obligated to critically "weed out materially incompatible commitments," to ampliatively "extract the material inferential consequences of each commitment," and to be "prepared to offer reasons for the commitments one acknowledges."[7] His Kantian account of personhood is governed by the idea of freedom that grounds the second *Critique*—the idea that "we are genuinely normatively constrained only by the rules . . . we adopt and acknowledge as binding on us."[8] However, had it been up to us not only to bind ourselves by the rules we adopt, but also to *make* them, then, as Wittgenstein famously argues, they would lack all normative force. In Brandom's account, Hegel's "social model of reciprocal recognition" compensates for this deficiency by setting the taking of responsibility and exercise of authority by individual agents in a social, intersubjective context in which agents hold each other responsible and acknowledge their authority.[9]

Brandom goes to greater lengths than any other modern thinker in situating the rationality of individual self-determining and deliberation within its broader social setting. Yet he remains largely oblivious to the dynamic, self-critical dimensions of human rationality, which he narrowly limits to testing commitments for mutual compatibility, as well as to the normative diversity characteristic of all but the most regimented of social phenomena.[10] As a result, the problem of holding one's commitments in normative check is not even registered, and the potential role that criticism from without may have to play in this regard goes unnoticed.

The way forward, I have argued, is to realize that, although we are incapable of normatively criticizing our own norms to any effect, or of being convinced of their normative failings by others, exposure to the normative criticism of others (not unlike exposure to a disturbing playback device) can at times sufficiently destabilize the commitments it challenges to render us ambivalent toward them. And commitments to which a person becomes ambivalent lose their normative hold and can be reassessed unproblematically in the light of his or her remaining norms.[11]

We are capable, in other words, of becoming ambivalent, but not of ambivalating *ourselves* toward our norms and standards. When we are left to our own devices, the self-critical fault lines of the reflective self are beyond our control and remain rigid and rationally unchangeable from within. But trusted criticism from without can sometimes change them for us. When the rich intrapersonal dialogue of the self is set within the context of an equally rich critical dialogue with others, Brandom's restricting neo-Hegelian corrective can be transcended, I argue, and effective transformative normative self-criticism can become a real option.

The application of these insights to science, in chapter 4, sets forth from two premises. First, because the sort of self-critical disposition needed for setting a paradigm shift in motion requires a challenging environment of potentially ambivalating normative criticism, such criticism will clearly not be forthcoming from within the paradigm-governed discursive settings of the field in question. The initial source of destabilization responsible for transforming a field must be external to it. Second, although a scientific field can only be ambivalated from without, it can only be *transformed* from within, by practitioners of voice and standing. Hence, the two main tasks faced by any account of scientific framework transitions that purports to be rational are, first, to discern the science's external sources of trusted normative criticism, and then to explain how and by whom their subsequent challenge was converted into an internal transformative force.

Accordingly, scientific paradigm shifts are described and analyzed in chapter 4, and exemplified in detail in the remaining chapters, as two-stage processes, in which reputed practitioners are first ambivalated by external critics, and later (inadvertently) succeed in creatively promulgating their newfound indecision within their communities by means of a particular form of published work. Both stages represent irreducible moments of intense intrasubjective deliberation that do not submit easily to the collectivist modes of scientific narration that have come to dominate science studies since Kuhn—hence, the notion of "scientific agency" alluded to in the book's subtitle.

To account for the first stage, I enlist, extend, and rework Peter Galison's novel notion of the scientific "trading zone"—the "locales" of professional engagement that practitioners frequent outside their home communities, as when bidding for grants, engaging students, or "trading" with neighbor disciplines for techniques or instruments. The point stressed is how often practitioners abroad are required in such settings to articulate and defend basics of their field that are passed over in silent agreement back home. It is here, I argue, that they are most liable to experience the potentially destabilizing and ambivalating effect of trusted external questioning.

The second stage is achieved when the doubts and indecision of such individuals come tacitly to inform their creative efforts to overcome them. Such efforts, it is argued, typically take the form of uneasily split, hybridic attempts to re-represent their field's basic assumptions by retaining some of the old while groping imaginatively toward new possibilities. When analyzed prospectively, such works can be seen to represent anxious, incongruous, yet highly creative departures from heartfelt commitments, capable of motivating others to seek a cleaner, more radical break. The profoundly split and discordant pictures presented in Tycho Brahe's planetary theory, Galileo's analysis of projectile motion, and Poincaré's geometrical conventionalism, for instance, all unwittingly preserved and, in so doing, powerfully propagated the keen ambivalence that begot them, prompting others to take a firmer stand.

My original plan for part 3 of the present study was to put the account of framework transitions developed in theory in part 2 to the test of a complex set of four related and interconnected early Victorian case studies on which I had worked in the past, and which I was eager to revisit with my new narratological framework in hand. These were the novel Janus-faced account of algebra presented in George Peacock's seminal *A Treatise on Algebra* of 1830, the differently hybridic views of mathematical physics evidenced in the writings of John Herschel and William Rowan Hamilton during the early 1830s, and William Whewell's highly original, yet similarly split historiography and philosophy of science published later in the decade. Each of the four, I submit, represented a profoundly ambivalated attempt on behalf of a profoundly creative thinker to come to terms with a significant external challenge to his heartfelt initial set of meta-scientific commitments. And each of them, in its way, helped set in motion a framework transition in the manner described. All four are briefly outlined at the outset of chapter 1.

However, as work proceeded, it swiftly became apparent that it was impossible to do justice to all four cases within the confines of a single book. I decided, therefore, to focus on Peacock, for several reasons. First, the story of his meta-mathematical work best exemplifies the theory the book seeks to promote, in its origins, its stark hybridity, and its impact on the leading mathematicians of his generation. Second, as we shall see, it relates to, and weaves in and out of, the other three stories. Third, the only other of the four stories that matches its depth and complexity is that of the formation, structure, and impact of Whewell's philosophy of science, and I felt that the work I had already published on Whewell covered that ground.[12] Fourth, we are less accustomed to speak of framework transitions in mathematics than in the natural sciences. And finally, the particular framework transition in question—from holding a "platonic" view of algebra as a truth-governed "science of number" or quantity (or as Kant and, later, William Rowan Hamilton had it, of pure time) to viewing it as a purely formal calculus of operations—remains a philosophically interesting, yet largely unappreciated chapter in the history of mathematics that, in my opinion, richly deserves to be told in detail.

The following chapters attempt to raise and answer a question that has exercised historians and philosophers of science for several decades. But they do so by appealing to a broader and more general array of philosophical discussions than is customary in the history and philosophy of science. There is an element of protest in thus orienting this study. For I believe that philosophy of science has not been sufficiently attentive to some of the good work done since the 1950s and 1960s in the philosophy of language and mind, and more recently in that of agency, self, normativity and rationality. As this study undertakes to show, there is much for the field to learn from these discussions and, conversely, much that scholars in those areas can learn from their application to understanding science. This is one sense in which this book about the ambivalating potential of neighboring fields with respect to one's own exemplifies itself. But it does so perhaps even more interestingly in the way it evolved in the course of my own particular vacillating between the history and the philosophy of science. I have chosen, therefore, to introduce it in chapter 1 by way of a semi-autobiographical charting of, and philosophical musing upon, its genesis and coming of age.

PART I

I (Orientations)

CHAPTER ONE

Two Beginnings: An Overture in the First Person

HISTORY AND ITS POPPERIAN PROBLEMS

In 1999 I published a bulky essay on the making of George Peacock's 1830 *Treatise on Algebra*, which described it as a work of "creative indecision" that convinced no one, yet played a crucially important role in the transformation of British mathematics during the 1830s and 1840s.[1] The Peacock paper was the last of a series of historical, interpretive works I had devoted during the 1980s and 1990s to early nineteenth-century British science and mathematics,[2] and it left me somewhat baffled and unsure as to their broader historiographical significance. I was then still firmly committed to a Popperian approach to the growth of knowledge, which I applied with a relish to the stories I was telling, and which at one point I even attempted to further elaborate philosophically.[3] Knowingly oblivious to the hold that the paradigms or conceptual frameworks to which the heroes of my stories were committed might have had on their thinking, I analyzed their works party-line Popper as imaginative attempts to solve the problems I believed they believed they were facing, constrained only by the limits of their creativity.

I focused on individuals—on their deliberations, on what they deemed wrong, and on their attempts, sustained or aborted, to put matters right—because I believed in the explanatory primacy of individual agency with respect to the knowledge claims they produced as well as to the communities, languages, institutions, and other collectivities in and with which they plied their trade. Such collectivities obviously determine an agent's setting, idiom, and point of departure, but I firmly believed that as the exclusive product of human endeavor, it is human endeavor that should ultimately explain their being and becoming, rather than the other way round. The individuals I studied—Whewell, Herschel, Babbage, Peacock, W. R. Hamilton, Maxwell, and others—all played major roles in dramatically changing the way the science and mathematics of their day was done and understood. All worked within and against the social, conceptual, normative, and institutional frameworks they were instrumental in building, modifying, or replacing. My interest lay in their *rational* impact—in exposing their reasons for adopting the perspectives they chose to adopt, for retaining the positions they chose to retain, for replacing those they chose to replace, and for entertaining those with which they chose to replace them. I was not interested in justifying their moves, only in understanding their own reasons for making them.[4]

These essays thus tacitly premised an internal-external divide that differed significantly from Lakatos's notorious model.[5] For Lakatos, the dividing line between a science's internal or rational component and its a-rational, external part ran between episodes in its history whose outcomes passed the test of what *the historian* considered to be the true method of science, and those that failed it. The rationality of scientific decision-making has nothing to do with the reasons or motivations that those who made the decisions had for making them, as long as what they actually achieved conforms retrospectively to the dictates of (the historian's pet) latter-day methodology.[6] On a Lakatosian showing, rationality is not an evaluative category of act*ing* or of agency, but a category of outcomes or of moves, of the products of acting, of the produce of agency judged by hindsight. A Lakatosian historian defines the internal-*cum*-rational component of a science in accord with what he, the historian, deems to have been the right decision to make and the appropriate move to make at the time. My work adopted a very different approach, which located the rational in what the actors of old, who made the decisions and made the moves it studied, deemed prospectively to be the right and appropriate thing for *them* to do. If Lakatos judges the rationality of others by the measure of his own norms and standards of approval, I attempted to judge them by the measure of the norms and standards of approval with which I considered them to have reasoned. The inner, rational component of scientific and meta-scientific endeavor was, and remains for me, the part deliberated and pursued prospectively and self-consciously by the actors in question, as opposed to the moves they made unthinkingly. Whether or not their actions won my own approval was and remains, *pace* Lakatos, quite beside the point.

The difficulty that began to present itself while I was writing the Peacock paper, and to which I have since devoted much of my writing and teaching, is that although the difference between willed and reasoned, and hence rational action, on the one hand, and caused, involuntary, and hence a-rational response, on the other, is categorical, the distinction I was implicitly advocating between them turned out to be far less clear than I had assumed. Although I was assessing the work and thinking of those I studied in the light of *their* norms and standards, I had at the time little appreciation and no understanding of the nature of the formative hold a person's normative outlook has on his or her capacity to reflect and reason. I failed to grasp the difficulty involved in accounting for the possibility of normatively criticizing and modifying a normative outlook from within—a problem from which the likes of Lakatos exempt themselves, and to which my avidly Popperian studies of the 1990s remained blissfully oblivious, even though the practitioners I was studying were doing exactly that!

By the time the Peacock paper was under way, I had begun to realize that Whewell's, Peacock's, Herschel's, and Hamilton's deliberations on the nature of science and mathematics were not easily accommodated by the simple processes of trial and error my Popperian background led me to expect. It was a far more tortured and convoluted process, one that ran up against the very basics of their respective worldviews, and hence of their deepest commitments. This realization was initially prompted, not by philosophical argument, but by the particular difficulties I encountered in narrating the particular stories I was trying to tell. Philosophical arguments came later and with considerable force, but they might not have had the same transformative effect on my thinking had my initial Popperian leanings not been first challenged by more purely historiographical issues—a point that nicely illustrates the philosophical positions I would eventually reach that constitute the second beginning of the present project. But I am getting ahead of myself.

In addition to the influence of Popper, I was impressed by Collingwood's "logic of question and answer."[7] My historical studies sought to retrace the erotetic trajectory of self-questioning by which their several heroes arrived, via various intermediate positions, at their final positions. This is never an easy task, but in my case it proved especially tricky, since none of them left any record of their actual pondering. The works they published, even when narrated in the first person, hardly ever adopted an autobiographical stance or even a reflective tone. They presented and argued for their *conclusions* rather than recapitulating the process by which they were reached. None of them kept diaries or log books of the kind that might have helped, and in their private correspondence, though typically voluminous, they rarely pondered aloud. So although my aim was to produce a forward-looking, prospective account of how their thinking developed, I was forced to work backward: Starting from the works they eventually published, working back through whatever unpublished drafts, notebooks, and letters I could lay hands on to their initial points of departure, speculating retrospectively as best I could about the problems to which they might have been responding and the questions they might have been asking themselves, as they made their way prospectively toward the culmination point of their efforts.

All historical work is to some extent marred by hindsight. Historians almost always know in advance the outcomes of the processes they are studying, and are almost always acquainted with the works eventually produced by the people whose deliberations they seek to reconstruct. Writing good history requires imaginatively *ignoring* any foreknowledge our subjects could not have had, attempting to recapture the essential open-endedness of their state of relative innocence with respect to the eventual outcomes of their efforts. All of which, as Collingwood urged repeatedly, requires the empathetic ability and willingness to see the world through *their* eyes as *they* pondered and deliberated how best to contend with the difficulties *they* believed *they* faced, still unsure of where exactly *they* were heading.

But, as in Collingwood's own archaeological examples,[8] in the absence of any direct evidence of my subjects' deliberations, ignoring the conclusions I knew they would eventually reach in favor of attending empathetically to their ponderings was simply not an option. Although I wholeheartedly conceded Collingwood's approach, the question I constantly found myself having to ask in these essays ran counter to the natural, lived, prospective direction of erotetic reasoning it prescribed. Instead of asking at each juncture why the person I was studying chose to answer the questions he faced in the particular way he did, I was constantly forced to ask what could have been the *questions* he was facing that might have prompted him to so answer them—a different question entirely! The former is the kind of question those I was studying were asking themselves all the time; the latter was one they would *never* have asked themselves. Reasoning from given question to possible answer is to attempt to *emulate* their thinking. But to reason back, as I was doing, from given answer to possible question is to adopt a decidedly external, outsider's view of one's subjects' reasoning quite alien to the deliberative path they were following.

Trying to follow my subjects' forward-looking deliberations, with my eyes constantly fixed to the rear-view mirror, as it were, certainly ran counter to the narrative sequence I sought to re-create, but it did have one important advantage. Backtracking from product to process prompted me to read my subjects' finished works as considered emerging possibilities rather than as confidently presented final conclusions; to imagine how they might have seemed to their authors just before their final endorsement. And as I did so, my attention was naturally drawn less to the details than to the more general form and structure of those works. It was then that I became aware of the odd yet telling feature they all shared: These works, all written during the 1830s, insisted on describing science and mathematics by splitting them down the middle and presenting them as rather unstable, hybrid amalgamations of two quite separate and conflicting undertakings.

Thus, in 1833, we find William Rowan Hamilton, celebrated Irish physicist and mathematician, insisting on there being two, rather than one "science of dynamics": "one subjective, *a priori*, metaphysical, deducible from meditation on our ideas of Power, Space, Time; the other objective, *a posteriori*, physical, discoverable by observation"—a view he was careful to distinguish from empiricist, Kantian, and what would later be dubbed instrumentalist accounts of the relation between the two. Of the former—the subjective, theoretical science of dynamics—he writes, "I account [it] indeed higher in dignity; but do not consider it as including the other, or as adequate ground to us for the expectation of any one appearance." Hamilton viewed the "wondrous convergence" of the two sciences as owing to "some mysterious union" residing in "the Divine Mind."[9]

Three years earlier, George Peacock's seminal *Treatise on Algebra* of 1830 had proposed splitting modern algebra into two awkwardly related, yet quite separate algebras, pertaining, as in Hamilton, to very different spheres of mathematical activity: on the one hand, "arithmetical algebra", conceived of as the science of number and its relations, duly devoid of such paradoxical notions as negative and imaginary quantities; on the other, "symbolical algebra," perceived as a wholly formal calculus of unconstrained symbols and operations, in which negatives and imaginaries and much more could find a natural place. Here also, as in Hamilton's case, no simple subsuming of the two was assumed. Peacock explicitly envisaged symbolical algebra *not* as a generalization of its arithmetical counterpart, nor did he view the latter as a mere application of the former to numbers. Peacock conceived of the two, as we shall see in some detail in due course, as the fruits of independent, yet curiously supplementary mathematical efforts.

In his extensive studies of the sciences conducted throughout the decade, William Whewell, Cambridge polymath and the leading historian and philosopher of science of his day, insisted, similarly to Hamilton though less crudely, and with considerably more philosophical finesse, that each of the "inductive sciences" comprised two integrated, mutually cultivated, yet antithetical components, arrived at by means of a two-pronged methodology geared to the attainment of two quite different notions of truth. The term *antithetical* is his. "The Fundamental Antithesis of Philosophy" is the title of the essay he first published in 1844 and later incorporated into the 1847 second edition of the *Philosophy*, in which the not-quite-Kantian epistemology grounding that work is set forth systematically.[10]

And to these explicit and acknowledged splittings presented in the works of Whewell, Hamilton, and Peacock, one should add John Herschel's meticulously compartmentalized writings on natural philosophy and mathematics, which, taken together, strongly imply a similar and even ruder split (though never explicitly presented as such) that segregated the factual and the formal components of mathematical physics to near-incongruous spheres of intellectual pursuit.

What was one to make of the fact that during the crucially formative years of their second coming of age, British science and mathematics were deemed insusceptible to unitary description by some of their keenest practitioners—a tendency that seems to have flourished briefly and noticeably in England during the 1830s and early 1840s, and then to have vanished almost as abruptly as it appeared? This is the larger question that gradually emerged from my historical work. All the dualisms in question were to some important extent strained and inherently unstable. Roughly speaking, one side of each of them pertained to or at least built upon a set of commitments from which its author originally set forth, while the other represented new and radically diverging possibilities.

All of this strongly suggested to me that, as innovative as they were, all these works bore the distinct mark of the kind of irresolvable inner struggle capable of yielding no more than a shaky compromise. They struck me more and more as plagued by profound, yet *inventive* indecision—as ingenious attempts to hold on to the old while being forced to grope creatively toward new options. But there was nothing fleeting or hesitant about them. The accounts of algebra and inductive science proposed respectively by Peacock and Whewell, like that implied by Herschel, represented serious, confident and detailed undertakings that took their authors years, if not decades, to develop, articulate, and refine. Hamilton never made public the explicit, twofold vision of dynamics he described privately to Whewell, yet it, too, was long in the making and strongly implied by his influential work both in optics and dynamics before turning his full attention to quaternions in the mid-1830s (in acknowledged response to Peacock's algebra).

Inventively undecided; creatively split; tortured, yet confident; reactionary, yet avant-garde—curiously, my own account of the works I was studying was beginning to resemble their own strained, hybrid structure! It was as if *I* lacked the vocabulary to properly assess them. Indeed, the Popperian-Collingwoodian vocabulary to which I was committed failed to do justice to these works. Rational agents, it implied, were expected to face up to the problems they encountered, to boldly address and solve them. An inability to fully relinquish past commitment in favor of less problematic options, it firmly implied, is a form of weakness, a lapse of rationality. The ideal coupling of keen and impartial refutation with bold and creative conjecture leaves no room and has little patience for the apparent dithering these works displayed.

And yet, the more I studied them and their authors, the more impressed I became with the dedication, the intensity of engagement, the sheer novelty, depth, and brilliance they displayed. But my evaluative vocabulary held me back.

Even more confusing was trying to do justice to their historical role—which again seemed paradoxically two-sided and at the same time quite incomprehensible from a Popperian perspective. On the one hand, despite their authors' status and standing, and despite being widely known and discussed, all of them swiftly vanished from the historical record without leaving a visible trace—at least in the accepted sense of the term. The hybrid depictions of science and mathematics proposed by the works of Whewell, Peacock, Hamilton, and Herschel all failed to generate a following. One would be hard pressed to name a single bona fide Whewellian philosopher of science,[11] a Peacockian algebraist or philosopher of mathematics, or anything like a Herschelian or Hamiltonian philosophical school of mathematical physics. And yet, each of these men commanded a major presence and played a centrally important role in the middle years of the radical reform of British science and mathematics during the first half of the nineteenth century.[12]

Here again, I found myself obliged to talk in near-oxymoronic terms of the crucially important role these works played despite their apparent total lack of impact, of how they failed to *influence* anyone, yet were centrally formative in the type of *reaction* they produced. I had obviously come up against a curiously complex historical moment that puzzlingly defied the philosophical and historiographical sensibilities with which I purported confidently to approach it. It was undoubtedly an extraordinarily pregnant moment in the course of a profound turning point in the life of British science and mathematics, but the problems it raised for me did not strike me as uniquely anomalous in any way. They pointed not to something out of the ordinary about the *objects* of my research, but to a serious failing in the way I was approaching them. This, I was beginning to understand, had to be seriously rethought. The rethinking process took the best of a decade and marks the second beginning of the present project.

RETHINKING PHILOSOPHY

Rationality and Criticism

I was then and remain today fully committed to the equation of rationality with criticism that grounds Popper's philosophy. The basic idea is simple enough. To act is to undertake to introduce some change in one's world or in one's view of it. To refrain from acting is to undertake to leave one's world or one's view of it *un*changed in the relevant respects. To do either rationally is to do so *for a reason*. And to either act or refrain from acting for a reason necessarily requires critically assessing the object of one's intended action or inaction. To have reason to act is to find the object of one's action sufficiently lacking to justify intervening; to have reason not to act is to find it preferable to the considered alternatives. Hence the crude rational force of problems and failings, which renders problem-seeking and problem-solving the hallmark of rationality for Popper, and provides the cornerstone to his entire system of thought.

But there is more to the identification of rationality and criticism than meets the eye. First, all criticism involves some element of *self*-critique, even when the immediate object of one's criticism is not one's own view of things. To find a state of affairs wanting or to find fault in someone else's view of things, is willy-nilly to find one's prior assessment of them to have been mistaken or incomplete. Awakening to a problem necessarily presupposes a measure of self-reckoning and displays a measure of self-correction. To take or to advocate taking action, as Popper states repeatedly, is hence in some irreducible sense to acknowledge and confront one's own fallibility.

However, while acquainting myself with the philosophical literature on agency and rationality more broadly, I awakened to the fact that a more basic element of critical self-scrutiny is inherently involved in the establishment of agency that far precedes criticizing others—a form of self-questioning that underlies our very capacity to reason. Talking about reflection, Christine Korsgaard makes the point vividly:

> Our capacity to turn our attention on our own mental activities is also a capacity to distance ourselves from them, and to call them into question. I perceive, and I find myself with a powerful impulse to believe. But

> I back up and bring that impulse into view and then I have a certain distance. Now the impulse doesn't dominate me and now I have a problem. Shall I believe? Is this perception really a *reason* to believe? I desire and I find myself with a powerful impulse to act. But I back up and bring that impulse into view and then I have a certain distance. Now the impulse doesn't dominate me and now I have a problem. Shall I act? Is this desire really a *reason* to act? The reflective mind cannot settle for perception and desire, not just as such. It needs a *reason*. Otherwise, at least as long at it reflects, it cannot commit itself or go forward.[13]

Self-criticism is inherently involved not only in *taking* reasoned action but in the more basic and prior process of reflectively *determining* one's reasons for acting. However, rationality and with it self-criticism are not confined to one's intrasubjective deliberation of reasons. There is a public and essentially intersubjective dimension to human rationality, in the light of which rationality's critical self-reflection can be seen to assert itself even more profoundly. This became apparent to me in the work of those who follow Wilfrid Sellars in identifying human rationality with "living," "being at home," "participating," "inhabiting," or "standing," as they variously have it, in "the space of reasons."[14] Sellars originally referred to it somewhat misleadingly as a "*logical*" space.[15] But given the inferential, and essentially communal, nature of the linguistic context it premises, the "space" in question clearly refers to a real discursive *agora* in which real fellow reasoners engage in "justifying what [they] say." In Brandom's more developed picture, agents are described as taking part in what he calls the game of giving and asking for reasons and keeping "deontic score" of each other's implied inferentially connected networks of commitments and entitlements. This, I felt, strongly suggests (though nowhere acknowledged by Brandom) a view of rationality that entails an even more radical element of self-doubting than Korsgaard's.

While Korsgaard locates the defining feature of rational agency in the essentially intrapersonal domain of deliberating which of our sensations, beliefs, and impulses merit endorsement as reasons, the thinkers of whom I speak (and of whom I take Brandom as representative) locate it in the essentially *interpersonal* domain of the public discourse of reasons. But they do not explain why.

Some would be inclined to play down the difference. For what is it to act *for* a reason in Korsgaard's sense, they might argue, if not to be willing, if prompted, to state one's reasons for so acting? Reflection, self-doubt, self-criticism, and the eventual reasoned decision to act or refrain from acting all belong, by definition, within the realm of the intrasubjective. The outer, intersubjective space of reasons, the argument would go, is simply where, when questioned, one's inner reasoning is aired and made explicit. But I seriously doubt such a position could be reasonably attributed to Sellars and Brandom. For them, to partake in the game of giving and asking for reasons is not the outward manifestation, but the defining feature of rationality. What ultimately makes us rational is for them not our capacity to *act* for reasons so much as our capacity and willingness to discuss them with one another. Playing the game of giving and asking for reasons is taken, on such a showing, not as an *indication* of rationality but as the very capacity that sets us apart as rational.[16]

Perhaps the difference indeed runs deeper, and has to do with these thinkers' linguistic pragmatism, in which the communal, social institution of language is accorded explanatory priority with respect to thinking. For those who understand thought as internalized speech, rather than vice versa, "what counts as a good reason depends on logical and conceptual-semantic rules that are followed intersubjectively. These can be read off the practices of a linguistic community," as Habermas puts it.[17] From Korsgaard's more strictly Kantian perspective, the intrasubjective discourse of self-scrutiny and self-government is basic, but in moving "from Kant to Hegel" (as Habermas aptly characterizes Brandom's *Making It Explicit*),[18] intersubjective discourse takes precedence, the argument would go, but without encroaching on, or in any way going substantially beyond, Korsgaardian reflective endorsement.[19]

But even if one grants the pragmatists their point, why should the order of pragmatic or linguistic explanation have anything to do with *defining* rationality? That an ability or willingness to join in the intersubjective discourse of reasons should be deemed (as indeed I think it should) a *necessary condition* for rational agency does not render it *constitutive* of rational agency. To *conduct oneself* rationally is in essence not to discuss or even to argue about reasons, but to act *for* a reason: to make the reasons one endorses one's motives for doing, thinking, or asserting one thing rather than another. Insofar as joining in the discourse of reasons is understood as no more than an *indispensable requisite* for rational action, the *definitional* status it is accorded for rationality by Brandom and others remains unexplained.

I realized that I was missing the point by attending to the wrong question. Since it is not disputed that rational agents do in fact engage in the game of giving and asking for reasons, and since by all counts doing so is an *activity*, the first question to those who deem doing so the epitome of rationality should be: *why* do we play the game? For what reason, or to what end? What could motivate rational agents to be curious about other people's reasons for acting and to be eager to expose their own? If doing so is accorded the definitive status for rationality it is claimed to have, doing so must at least *be* rational, that is, performed for a reason!

None of the thinkers mentioned raises the question, nor do their writings offer much implicitly by way of answering it—with the exception of a single, declarative, yet seminal text of McDowell's to be discussed below.

The pragmatist order of explanation aside, it would be quite wrong, it seemed to me, to view the interpersonal activity of giving and asking for reasons simply as an externalization of Korsgaardian intrapersonal, reflective deliberation. Hers is not a discourse *of* reasons, but a critical scrutinizing of sensation, desire, and impulse *for* reasons. Put crudely, what Korsgaard describes is the initial reflective process of determining whether what at any time one wants is what one really *wants to* want, as Frankfurt would put it; whether what one would like to think and do is what one believes one *ought* to be thinking or doing. It is not to play the game of giving and asking for reasons in private, so much as it is a prerequisite to playing it in public. One must have reasons to be able and willing to give them, and one wouldn't ask others for theirs if one didn't think they already had them.

Yet—and this is my main point—although obviously ordered, it would be equally wrong to view the two processes as distinctly separated, as if one's reasons had to be fully worked out and finalized in private before being aired publicly. If that were so, what would be the point of making them public? Had we been capable of achieving confident closure regarding our reasons in the course of the Korsgaardian, intrasubjective phase, what would be the point of discussing them with others? Vast areas of our inner life remain private—beliefs, feelings, impressions, and judgments that most of us would never think of sharing with others by making them public. And yet we do engage insistently in the game of explaining our reasons to others and asking them to explain theirs. To some extent we do so in order to *better understand* our fellow reasoners. Knowing their reasons for action and inferring and keeping score of their other implied commitments and entitlements enhances our ability to predict how they will respond in various circumstances and to better calculate our own response to our best interests. For the sake of brevity, call this the game-theoretical account. It is obviously part of the story, on a par with various other kinds of practical knowledge and knowhow we master and manipulate to navigate our surroundings in pursuit of our goals. But it cannot be the whole story for those who consider the public discourse of reasons definitive of human rationality.

For one thing, the game-theoretic account can explain the incentive for (or, rather, the utility of) asking, but less so that for *giving*, one's reasons for acting, except perhaps as a trade-off. Second, from a purely game-theoretic perspective, Brandomian scorekeeping is just one more means among many that agents instrumentally employ to better facilitate their ends—in this case, by helping to factor in the human element of their environment. In this respect it is not very different from, and in many contexts less important than, acquainting themselves with other relatively predictable aspects of their world. Hence, apart from the fact that such scorekeeping can only properly explain one aspect of the discourse of reasons, it remains quite unclear, from a game-theoretical perspective, what could possibly merit singling it out as the quintessential embodiment of rationality.

All the more so when viewed from the perspective of the thinkers in question. None of them *deny* the rationality of instrumental reasoning, but neither do any of them consider striving deliberately to get what one wants as exhaustive or in any way definitive of rationality. For Brandom, for instance, instrumental rationality thus construed is accorded only second place in the ascending, nested list of five "conceptions of rationality" discussed at the outset of his *Tales of the Mighty Dead*.[20] Playing the game of giving and asking for reasons, I therefore concluded, serves a very different and far more basic, if largely implicit, purpose in the work of Sellars, McDowell, and especially Brandom, which has more to do with *setting* one's objectives than with improving the means for attaining them. The function of the game is in an important sense *self*-exploratory. It is geared at *self*-understanding by engaging fellow rationals. It is best viewed, I believe, as the intersubjective *continuation, development, and follow-up* of the intrasubjective processes of reflective endorsement described by Korsgaard. At the outset of his most recent work, Brandom makes the point I am driving at almost explicitly:

> To be a rational being . . . is to be subject to a distinctive kind of normative *appraisal*: *assessment* of the *reasons* for what one does—in the sense of 'doing' that is marked off by its *liability* to just that sort of appraisal. Rational beings are ones that *ought* to have reasons for what they do, and *ought* to act as they have reason to. They are subjects of rational obligations, prohibitions, and permissions.[21]

What is gained by engaging other rational agents in giving and asking for reasons is the kind of *critical input* one is not very good at giving oneself. Schematically speaking, the initial stage is intrasubjective—Kantian, if you like. We engage others as rational beings after having first subjected ourselves as best we can to the kind of normative self-appraisal of which Korsgaard speaks. But by going on from there to expose our reasons to others, and to expose ourselves to theirs, we are subjecting our provisional judgments to their critical assessment and exposing ourselves to the reasoned alternative fruits of their reasoning. We thus enable our reasons—that is, what we had intrasubjectively taken our reasons to be—to be normatively assessed from a diversity of perspectives, while in doing so, we offer ourselves reciprocally as their normative appraisers.

And yet, clear as it is, Brandom's text waffles tellingly between two different senses of subjection to the normative appraisal of others: on the one hand, being held liable to *having* reasons for what we do and acting as we have reason to act, and on the other, having our *reasons* themselves normatively assessed,. Being held responsible for having any reasons at all to act, and for acting in accord with them, is to be held accountable for making ourselves agents in Korsgaard's sense of the term (or for "making up our minds," in Frankfurt's delightfully literal use of the term),[22] and for conducting ourselves consistently. This first form of normative appraisal amounts, one could say, to putting an intersubjective check on one's Korsgaardian, intrasubjective responsibilities for assuming human agency.

The second form of appraisal, I realized, consists far more radically in holding one accountable not merely for having reasons act, but for the *particular* reasons one has adopted for acting. What is questioned is not *whether* one has assumed agency, but the normative *merit* of the agency one has assumed. Playing the game to this second end is to play it very differently than to the first. To engage others in giving and asking for reasons merely for the sake of assuring them of one's own agency, and running a check on theirs, implies a keen exchange of information, but one that lacks *dialogical* and *normative* force. When asking for a person's reasons, any answer short of "No special reason, I just felt like doing it" or "It just felt right" will suffice, and end the conversation. Apart from its obvious instrumental value, keeping Brandomian score of the various answers garnered allows one, in addition, to hold the consistency and coherence of other people's reasoning in check, but little more than that. But if the game is conceived dialogically, as critically challenging the normative *worth*, rather than the mere *coherence*, of one another's normative choices, it becomes an invaluable intersubjective extension of the intrasubjective processes of reflective self-questioning, from which we each initially set forth as self-constituted agents, to use Korsgaard's most recent terminology.[23]

If this is what motivates the game, then the definitive role for rational action assigned to its intersubjective phase acquires depth, meaning, and significance. If playing the game of giving and asking for reasons is something one does rationally (or is at least something that bespeaks or manifests one's rationality), then it is not something that simply happens to rational agents, but something they contrive actively and knowingly to perform, an activity in which they *have reason* to partake. And the only conceivable reason for wanting to expose one's reasons to the normative critique of other reasoners is that one is in some way *unsure* about one's own judgments, and seeks, in doing so, to affirm or improve them.

Korsgaard, I said, is right in insisting that self-constitution belongs wholly and exclusively within the realm of the intrasubjective. No one but oneself can endorse one's reasons to act, and commit oneself to norms. No one but oneself can oppose certain of one's desires and impulses, and identify with others. To adopt the reflective stance is to stand in critical self-judgment, wary of the persisting claims of our sensations, desires, and impulses, just as Korsgaard describes. Having decided upon our reasons, we are in a position rationally to exercise our agency in action, which, as noted, is to adopt a similarly self-critical stance, not toward our desires and impulses, but toward our world and especially our picture of it. Constituting ourselves as agents and exercising our agency are hence profoundly self-adjudicating critical undertakings.

And on the reading at which I arrived, to engage in the intersubjective dialogue of reasons is, in an important sense, to extend the self-critical attitude that so thoroughly permeates our rationality by transcending it *reflectively*. Engagement in the dialogue of reasons demonstrates a second-order wariness and adoption of a self-questioning stance toward our first-order Korsgaardian self-critical choices. Juxtaposing the three components of human agency—endorsing reasons, applying them in action, and exposing them to the appraisal of others—our rationality emerges as so thoroughly steeped in self-criticism as to fully justify Popper and his followers in identifying the two.

On this view, it also became clear to me why Brandom and others were fully justified in viewing the third component—participation in the intersubjective exchange of reasons—as the epitome of our rationality. First-order rational engagement comprises a twofold, reflective exercise of our critical capacities—first, inwardly, toward our desires and impulses, and then, outwardly, yet nonetheless reflectively, toward the world as we see it. To subject ourselves to the normative appraisal of fellow reasoners is to apply our rationality to itself—as it were, to employ others in reflecting critically upon our own attempts at critical reflection, acknowledging our own fallibility by engaging their help. This is human rationality at its utmost, at the very point in which critical reflection comes full circle and critically engages itself. And it is precisely here, maintain Sellars, McDowell, and Brandom, that the "I" becomes "we," as the self awakens to its thoroughly social and discursive embeddedness. But this embeddedness is far from simple. And in probing its vexingly problematic nature lies the second, and philosophical, beginning of the present study.

Normativity and the Problem of Self-Criticism

Popper speaks much of "problem situations" as what rationally motivates taking action, implying that problems are objective *facts*—states of affairs that compel the rational to react by the sheer force of their disturbing presence. But the sort of problem that moves us to action is a state of affairs of a very particular kind: a situation that those whom it moves conceive as *undesirable*, as a *shortcoming*, as *wanting* to the point of *meriting correction*. A problem is a failing we deem *ought not* to be tolerated. All of these are decisively *normative* categories that necessarily presuppose and are defined by a developed normative outlook, or system of values and standards. In this sense a problem may well be considered a fact, but it is a decidedly value-laden, framework-dependent fact—a fact constituted by, and therefore necessarily relative to, a normative point of view.[24]

Since rationality presupposes criticism, and criticizing amounts minimally[25] to deeming a state of affairs problematic, rationality itself is rendered inherently framework dependent. At what I have been calling its Korsgaardian level of self-constitution, the level of selecting and endorsing reasons and applying them in action, the normative dependency of rationality is understandable and seemingly unproblematic. Deeming a desire, sensation, or impulse worthy or unworthy of being endorsed as a reason, and a state of affairs as in need or not in need of correction, are clear cases of norm application—norms to which one must necessarily be already firmly committed. Conducting oneself rationally at this level is to hold oneself and one's surroundings *accountable to* one's *existing* standards of propriety.

But, as we have seen, our rationality consists of more. Doing our utmost to do what we know is right is only part of the story. Rationality does not stop here, but extends beyond its Korsgaardian level in requiring us not only to apply our standards *in* critical appraising, but to subject them *to* critical appraisal! Rationality demands not only to act in accord with the standards and norms to which we are committed, but to hold those very norms and standards in normative check. Indeed, doing so in the company of others, insist the thinkers from whose work the present study take its cue, is what marks us off as rational. Korsgaard largely ignores the intersubjective setting of so doing, yet nonetheless makes the point forcefully with respect to the conceptual schemes of science. Substitute "norms" or "standards" for "concepts," and her text says it all:

> The normative question is a question about the status of the concepts, not about whether they have been correctly applied. Is our conceptual scheme adequate? Is it the correct one, or the best one, or the one that captures the most, or the one that captures what is "really" true about the world? Philosophers will of course disagree on whether any of these questions are coherent and, if so, which one is the right one to ask. But since science leads us to modify our conceptual scheme, and we think of these modifications as improvements, it does appear that some such question is in order.[26]

But how *can* such a question be in order? How *can* it be coherent? How can the standards (in this case, the conceptual framework or scheme) to which we are committed be effectively subjected to rational appraisal if it is by means of those very standards that we rationally appraise? How can one be expected to take action rationally against the very basis of one's rational judgment? Brandom implies, following Hegel, that this is precisely what exposing our reasons and actions to others is supposed to achieve (an element wholly lacking from Korsgaard's more strictly Kantian discussion). But this, too, is unclear. If a person is incapable of critically appraising his very standards of appraisal—except in the trivial, and quite unhelpful, sense of circularly *approving* them—how could he be convinced to do so by others? If a normative system cannot be found normatively wanting from its own point of view, what can be achieved by exposing it to other points of view? Arguing on the basis of normative commitments different from our own would be irrelevant, while arguing from ours seems absurd! It is one thing to doubt, or to be wary about, one's commitments, but quite another to fashion a normative argument against them. To merely wonder whether our conceptual scheme is the right one or worry that it is not, and to be moved by argument to replace it, are two quite different things. In addressing "the normative question" rather than the rational exercise of agency, Korsgaard's text conflates the two. McDowell, on the other hand, in a passage I have analyzed in some detail elsewhere,[27] raises the same issue with respect to ethics far more pointedly:

> Like any thinking, ethical thinking is under a standing obligation to reflect about and criticize the standards by which, at any time, it takes itself to be governed. . . . Now it is a key point that for such reflective criticism, the appropriate image is Neurath's, in which a sailor overhauls his ship while it is afloat. This does not mean such reflection cannot be radical. One can find oneself called on to jettison parts of one's inherited ways of thinking; and though this is harder to place in Neurath's image, weaknesses that reflection discloses in inherited ways of thinking can dictate the formation of new concepts and conceptions. But the essential thing is that one can reflect only from the midst of the way of thinking one is reflecting about. So if one entertains the thought that bringing one's current ethical outlook to bear on a situation alerts one to demands that are real, one need not be envisaging any sort of validation other than a Neurathian one. The thought is that this application of one's ethical outlook would stand up to the outlook's own reflective scrutiny.[28]

McDowell zigzags uneasily between his desire, on the one hand, to see normative self-criticism self-applied so radically as to justify the replacement of major components of an ethical outlook, and the worrying thought, on the other, that because one can only reflect "from the midst of the way of thinking one is reflecting about," the most radical form of normative criticism available to an ethical outlook is its self-exposure, Neurath-style, to its "own reflective scrutiny."

McDowell and Korsgaard are in full agreement as to both the self-critical nature of rationality and its being an inherently normative undertaking. They both firmly insist that it is an undertaking that centrally obliges us not only to apply our standards critically, but also to apply reflective criticism *to* our standards; not to deem our standards good because they are ours, but in some sense to endorse them as ours because they are good. They are also both clearly aware of how problematic such a demand is. Korsgaard openly admits that many will find it incoherent, but refrains from offering a solution of her own. In *Mind and World* McDowell does not acknowledge the problem explicitly, but the way his argument there oscillates around Neurath's metaphor leaves no doubt that he, too, sees no clear way to accommodate it. In an essay he published a year later he is much clearer, however. Normative commitment, he argues, is a matter of upbringing—"a process of habituation into suitable modes of behavior, inextricably bound up with the inculcation of suitable related modes of thought." The only way that someone lacking such an upbringing, "someone who has in some radical way slipped through the net," could be "made to consider the relevant matters aright," he states, is by "something like conversion." The idea that such a person could be "induced into seeing things straight by directing some piece of reasoning at him" McDowell deems "massively implausible"![29]

Norms may, of course, be criticized in the light of higher norms—which is what Neurath's metaphor is about. We do it all the time. One can criticize oneself and others for failing to live up to, say, the standards of one's college, and quite consistently criticize those very standards in the light of higher (more fundamental) norms, such as those of one's religion, or one's liberal commitments. But there is a deep sense in which, taken whole, cases like these only prove that the lower-level (in this case, the college) norm was not really among one's normative commitments to begin with. Normative purging in the light of higher-level commitments is how we find out what we are, in fact, committed to, not whether what we are committed to merits commitment.[30] Standards of propriety to which we are genuinely committed, it would seem, are by definition immune to normative criticism. Hence one of two: Either we can claim that our *real* standards—namely, the norms to which we are *really* committed—are absolute and unchangeable, and therefore are, in principle, not subject to rational assessment and modification, or we can hold that standards do indeed change, but, in principle, can only do so surreptitiously, involuntarily. With regard to science, Popper would seem to choose the first option, in limiting the very notion of scientific problem to predictive failure, and ignoring all other normative aspects of scientific thinking and practice—including the constitutive, normative role of conceptual frameworks in science, raised briefly by Korsgaard, which, following Kuhn and Friedman, will be made central to the argument of the present study.

I doubt that Popper is justified in regarding predictive failure as both an absolute scientific standard and the only one that merits rational response. Kuhn also toyed with the idea that certain instrumental criteria, such as accuracy, precision, scope, simplicity, fruitfulness, consistency, and so on, are "necessarily permanent" across major scientific upheavals, and are somehow definitive of scientific success.[31] Yet neither Kuhn nor Popper is an instrumentalist. Both insist, albeit very differently, that the aim of scientific research is truth. Kuhn deems truth to be framework dependent;[32] Popper does not, insisting on a framework-independent, realist notion of truth by correspondence, which somehow transcends merely instrumentalist criteria as well as the normative hold of the scientific vocabulary and practices that practitioners internalize during their training and professional discourse.[33] The idea, centrally implied by Popper's philosophy, that the standards and norms embedded in one's language and practices can function as mere hypothetical media that creative individuals are able to stand back from, consider critically, and replace with alternatives that can be deemed in some objective, non–linguistically dependent sense better than their predecessors, is an idea wholly foreign to the approach adopted in the following pages. Rejecting it in favor of a view of normativity and a philosophy of language closer to Kuhn's marks the philosophical journey that constituted the second beginning of the present study.

History's Philosophical Challenge

As noted at the outset, what initially broke the hold of Popper's philosophy on my thinking, in favor of a position closer to Kuhn's, was the realization that the early Victorian works I was studying focused almost exclusively on second-order, meta-scientific, and meta-mathematical questions that bore decisively on first-order research, but with regard to which, of themselves, Popperian fallibilism seemed quite irrelevant.

Take Peacock's predicament, for example: Was algebra to be understood as a logically connected system of *truths* about the realm of number or of quantity, as he was taught and had always believed, or was it to be considered a purely formal, symbolical system, a logically structured *system of formal rules*, like those of a game, or of grammar, capable of achieving high levels of coherence, clarity, and completeness, but which were true *of* nothing? The two very different views of algebra generate very different ways of doing mathematics and very different, if not incompatible, notions of mathematical meaning and validity. And yet, they cannot be adjudicated by testing. For they were, for Peacock, not attempts at faithfully *representing* algebra as it was pursued, or as it was understood by its leading practitioners, or by the majority of its practitioners. Nor were they considered by him as ways to improve algebra's capacity to attain its objectives. Unlike scientific attempts to faithfully depict or represent, or effectively model, an external state of affairs, or theories designed instrumentally to effectively predict certain of their features, unlike plans or strategies or any other device geared to better, or more fully, achieve given objectives, the two views of algebra were in no clear sense considered by him as *means* to ends. They could not, therefore, be weighed, compared, or tested by any form of instrumental reasoning against a shared comparative criterion of success or failure. The two conceptions of algebra were for Peacock attempts to state how algebra *ought* be understood and pursued. They offered different *final goals* for understanding and pursuing algebra, if you wish, not different means of attaining them.

In short, they were *normative* alternatives, and as such defied any clear form of Popperian appraisal and adjudication. Each of them gives rise, of course, to partially overlapping complex notions of potential algebraic improvements and advances, failings and difficulties. On each of the two showings, algebra, as it stood, could differently harbor a complexity of shortcomings and flaws: open problems, unresolved dilemmas, unproven theorems, seeming and real inconsistencies, unintuitive or otherwise puzzling results, and so forth—problems of the kind algebraists grapple with daily. But there is no clear sense in which *they themselves* could be deemed problematic—except to be dismissed as wrongheaded from the perspective of the other or from that of a third normative alternative. There seemed to me to be no clear way in which either view of algebra could be deemed problematic from within by those committed to it, or considered inferior to the other from some external "objective" viewpoint acceptable to both. In the language I became accustomed to only later, they represented different mathematical forms of life, different paradigms, different conceptual frameworks constitutive of incompatible worlds of mathematical practice and mathematical evaluation.[34]

Peacock's *Treatise on Algebra* was not an idle exercise. Nor was it a case of detached philosophical musing. Neither were the two options a fork that somehow opened before him on the road he was taking. The realist approach that viewed algebra as "a science"[35] was the one to which he had initially been firmly committed. The second, formalist option was a view he encountered and initially opposed (with respect to algebra) early in the game—first in the approach to the calculus he shared with Herschel (but was reluctant to extend to algebra), and more significantly, in a bold and quite general unpublished work by Charles Babbage on the "Philosophy of Analysis,"[36] which Peacock had read and disapproved of during the early 1820s.[37] And yet Peacock ended up dithering anxiously between the two incompatible options, painfully straddling the unbridgeable gulf between them, having somehow become committed to both, yet singularly committed to neither, investing the best of his considerable mathematical knowledge and talents to fashion a hybrid combination of the two that was as ingenious as it was impossible.

His was not a gestalt-type conversion from one worldview to another, as Kuhn occasionally describes paradigm shifts.[38] Had that been the case, the resulting work would have been thoroughly and exclusively formalistic. It was not. Peacock remained committed to his former view. But he had somehow become sufficiently ambivalent toward it to seriously consider its rival, to which he became sufficiently devoted to be able to endorse the two views side by side in uneasy conjunction. He seems also to have clearly realized that the two could not be played off against each other, that they were "incommensurable" in the relevant respects.[39] Peacock's case seemed to fly in the face of both Popperian *and* Kuhnian accounts of major scientific change, occupying a middle position neither philosophy seemed capable of accommodating.

And the same applies to Herschel's, Hamilton,'s and Whewell's different meta-scientific writings. Herschel was from the start a staunch follower of Lagrange's thoroughly formalistic approach to the calculus, which he extended in time to mathematics in general.[40] Yet, quite unlike Lagrange, whose approach to mechanics was as formalistic as his version of the calculus, Herschel adopted a thoroughly realistic, Baconian view of physics.[41]

How was one to understand the combination of the two that created modern mathematical physics? What role had mathematics to play in modern physics, if it was not considered to be true of the real physical quantities to which it was applied? What could be gained by so mathematizing scientific knowledge from a Baconian perspective?[42] This is the main question that plagued Herschel's meta-scientific musings throughout his career. Unlike Peacock, Hamilton, or Whewell, Herschel kept his deliberations to himself. He made no move toward articulating the dilemma, or toward resolving it by showing how the two jarring commitments could somehow be juxtaposed. He chose instead to avoid the issue by relegating his writings on the nature of mathematics and on the philosophy of what he carefully called "natural philosophy" to quite separate publications. The fact that he remained faithful to both positions throughout his long career clearly attests to the depth of his commitment to them; the fact that he made no attempt to combine them, or even to raise the question of doing so, can be seen as proof of the depth of his ambivalence. Here again was one of the most creative, respected, and learned individuals of the era caught between painfully jarring normative meta-scientific options to the point of speechlessness!

Hamilton's quandary also concerned mathematical physics, to which he contributed famously. Hamilton, who subscribed to a realist (or, as we would say today, Platonist) rather than a formalist conception of mathematics, was not troubled, as was Herschel, by the interface between the mathematical and natural philosophical constituents of modern physics. His concern had to do with the dividing line between what we now call theoretical and experimental physics. He believed that the science of dynamics was developed and perfected, discovered and confirmed in the course of two seemingly independent and unrelated processes: one geared to articulating the science's conceptual scheme in theory, the other to painting a picture of the observed empirical facts, He viewed the former as cultivated by intense, inward-gazing conceptual analysis that yielded much more than the science's mathematical apparatus. The entire conceptual structure of dynamics—its notions of cause and force, velocity and acceleration, energy and momentum, along with their exact and complex relationships—was the exclusive creation of the mind, the combined product of intense *contemplation*, which, he believed, at no stage depended on experience. This notion of "the subjective, metaphysical" of the two sciences of dynamics owed to Hamilton's deep and early commitment to a romantic view of poetic insight and cognition, born of his significant early relationship first with William Wordsworth and then with Samuel Taylor Coleridge, which, once he had finally settled (not without deliberation) on his scientific vocation, he carried into science and mathematics.[43] And yet, calculated experiment and observation gave rise to a rich and objective picture of physical and astronomical reality of their own, which, he believed, they yielded quite independently of the theoretician's efforts.

The two scientific enterprises seemed to Hamilton to have nothing in common. They were motivated and informed by what we would today call quite different paradigms and generated by quite different scientific methodologies in the hands of quite different research communities. The materials on which they built were as different as the research methods they employed, the training each required, and the fruits they were each expected to bear. And yet, in some mysterious way the keen, systematic, inward-looking investigation of our conceptions and ideas ends up coinciding perfectly with the equally keen and systematic outward-looking investigation of empirical experience. In terms of his personal vocation, Hamilton obviously sided with the theoreticians. But his quandary was not about the kind of physics he himself should pursue. It was, again, a philosophical or meta-scientific inability to reconcile incompatible commitments regarding the means of *production* of modern physics, if you wish, and consequently, its source of validity. Unlike Peacock and Whewell, Hamilton had little to offer by way of bridging the divide. From any human perspective—that of an external observer or that of those responsible for cultivating and advancing the two projects—he deemed their convergence a coincidence of truly colossal dimensions, which he ultimately ascribed, as did Plato (endorsing something akin to Plato's ontology, but not his epistemology), to their mystical unity in the mind of God. Hence, there was not one science of dynamics, but rather two, uneasily and puzzlingly poised in relation to one another, converging perfectly by virtue of nothing except a divinely mandated, pre-established harmony! Like Herschel's thunderous silence, Hamilton's resort to mysticism bears dramatic witness both to the depth of his two wholly diverse sets of meta-scientific commitments and to the extent to which he found them incompatible.

Whewell's "antithetical" philosophy of science was the product of a meta-scientific journey more philosophically complex, more uneven, and more profound than those of his three close friends. As in the case of Herschel, the hybrid synthesis he ended up fashioning preserved, on its one side, an initial native commitment to empirical realism. The inductive sciences, he maintained throughout, were cultivated with the aim of corresponding to external reality. Developing a branch of science to perfection could not ensure that it would turn out to be true, and yet, those that did (his two paradigmatic examples were Newton's universal gravity and Huygens's and Fresnel's undulatory theory of light),[44] he insisted, could be recognized and warranted as such.

At the same time, however, he had become just as fully committed to a form of idealism that harked back to Kant, one that was far more radical than Hamilton's and according to which *all* the conceptual content manifest in human knowledge originates in the mind. All knowledge, all the way down to our most basic and immediate empirical observations, originates in acts of reading *in* rather than reading *out*. New knowledge, he insisted, is not discovered *in* the facts one considers—that is, gleaned or read *off*, or *out of* them, as empiricists would have it—but is rather *made* by superimposing on the facts a new "colligating" concept, by which they come to be seen in a new light, as he put it. Whewell's epistemology and philosophy of science were in part thoroughly Kantian (even containing interesting parallels to Kant's notion of synthetic a priori truth and use of transcendental reasoning),[45] yet, at the same time, they somehow preserved a determinately non-Kantian, empiricist commitment to the idea that the very best of science can be, and on occasion has been, proven true not merely of the world we experience, but of the world-in-itself![46]

Of course, fully developed Kantian and empiricist epistemologies and philosophies of science were available to Whewell, but he would settle for neither one exclusively. Nor was he willing to somehow retain them both separately, as Hamilton's proposal could imply. Whewell insisted that the two deeply incompatible, well-developed accounts of the acquisition, the development, and the nature of the truth of science be combined to form a complex philosophical hybrid quite unlike anything that went before or came after it. And again, as in the cases of his three friends, Whewell's most impressive five volumes of the history and the philosophy of the inductive sciences, notwithstanding the confident, even majestic scale and tone of the work, were born of nearly three decades of grappling painfully with conflicting commitments, strikingly preserving the ambivalence that begot them, and which they purported to resolve.

The more I learned about the four unstable, yet so insistent, split meta-scientific pictures, the less sure I became of how to make historical and philosophical sense of them. As noted above, closely studied and taken in union, the hybrid philosophies fashioned by Hamilton, Herschel, Peacock, and Whewell flew in the face of the Popperian philosophy of science I sided with, yet at the same time defied its Kuhnian rival: the former seemed oblivious to the function and the normative hold of meta-scientific commitments, the latter, though well aware of them, seemed incapable of accounting for the complex dynamics of rational, deliberated meta-scientific realignment.

THE VIEW FROM WITHIN

It was at this point that I became acquainted with the work of a young Israeli ethicist, Yitzhak Benbaji, whose doctoral dissertation I had the good fortune of reviewing. In it he forcefully raised and articulated in the field of meta-ethics the problem I had begun to realize I was struggling with in my own work.[47] I proposed that we join forces to collaborate in writing a book on the problem of rational normative self-criticism in general that would go beyond ethics and science by addressing it with respect to normativity in general.

The resulting work, *The View from Within: Normativity and the Limits of Self-Criticism*, took much longer to write than we had envisaged and was published by the University of Notre Dame Press ten years later. For one thing, when embarking on the project, I had still to fully appreciate the constitutive force of normative commitment with regard to belief and perception, and the depth of its hold. I was still toying with Agassi's idea[48] that a normative framework could be treated provisionally and instrumentally, and be normatively faulted in the light of its independently perceived "failures of practice," the term I was using then for failures of application. It took time for me to realize that one's normative outlook is not *applied* to normatively neutral and independently perceived states of affairs, as some perspectivists would have it, but is *constitutive* of their very perception.

The *View from Within* marks the second beginning from which the present study sets forth, and lays the philosophical foundations for the account of scientific framework transitions it purports to articulate and exemplify.

The large question it addresses is to what extent people can be said to change their normative commitments rationally. The problem is by now familiar enough: To claim that norms and standards can be changed for a reason is to claim that they can be normatively faulted by those whose speech and thinking they govern—a problem widely considered insurmountable. As much as we would like to follow McDowell and Korsgaard in demanding that rational agents not only exercise their norms and standards *in* self-criticism but make them the *object* of such criticism, no philosophical account of personal identity makes room for such feats of self-negation. For, as the argument is made explicitly by Wittgenstein, Rorty, and Davidson,[49] and implied by many others, it is impossible for one to coherently deem wrong one's very standards of right and wrong (as opposed to merely finding them inconsistent or unclear), or, for the same reason, to be convinced to do so by others. We agree: As a purely intrasubjective feat of self-evaluation, normative self-criticism is impossible. A person is incapable of forming such an argument that she could deem valid, or of being convinced of it by others. The book does not contest these claims, and yet it insists—and this is its main point—that normative commitment can nonetheless be sufficiently *destabilized* by *external* criticism to demand rational reconsideration. Although it is impossible to *convince* a person by normatively criticizing her norms, such criticism, we show, can have a decidedly, and transformative, *ambivalating* effect.

Amalgamating a Brandomian approach to normative discourse, Walzer's view of social criticism, and an account of personal identity that takes its cue from Korsgaard and especially Harry Frankfurt yielded a phenomenology of critical engagement that enabled us to make our case. "Amalgamating" is perhaps the wrong word to use. We certainly engaged their work and built on it, but merely conjoining their efforts was to no avail. Even when taken in union, their works, in and of themselves, yielded little by way of addressing the question we set ourselves. Brandom, Walzer, and Friedman, whose writings centrally address the normativity, rationality, and, in Walzer's case, the decisively *critical* nature of transformative discourse,[50] limit themselves exclusively to the intersubjective communal realm, showing no interest and offering no resources for understanding the possible transformative impact of those discourses on individual commitment and intrasubjective deliberation. The assumption seems to be, as one Israeli philosopher put it to me in a public debate, that intrasubjective deliberation is no more than internalized intersubjective discourse—which, of course, wholly ignores the normative hold a person's commitments can have on his very ability to reason.

Conversely, thinkers like the early Charles Taylor,[51] Christine Korsgaard,[52] and especially Harry Frankfurt,[53] who offer rich accounts of agency, selfhood, and reflective intrasubjective normative deliberation, show no interest at all in exploring the possible transformative impact that the normative critique of others might have on the selves they study.

Nothing is achieved by simply juxtaposing the two literatures: They simply pass each other by like ships in the night, without as much as a glance. Amalgamating them required us to return to basics and draw up a fairly detailed phenomenology of the critical speech-act and an analysis of the argument form it typically takes, and to do so with special reference to normative criticism—while endorsing and utilizing the best that the two literatures had to offer. The fruits of our efforts can be briefly summarized as follows:

- Criticism is more than a negative test result, it is a form of rebuke. Systems or objects can be tested, but only agents can be criticized. Criticism purports to hold its addressees *responsible* by *proving* that something is amiss that requires their attention. Unlike mere testing, it can therefore achieve its intended effect only when internalized as *self*-criticism. Hence, to be effective, critics must argue empathetically from premises they believe their addressees hold true.
- In the case of *normative* criticism, however, this is never an option, for there exist no premises to which one's addressees are liable to agree that entail a denunciation of their very norms. Because normative criticism cannot be endorsed and internalized, it can never *convince*.
- Sensing this, normative critics tend to argue from premises *close* to what their addressees hold true, *yet sufficiently different* to allow their arguments to stick. And thus, in the process of questioning our norms, normative critics typically convey to us a portrayal of our relevant commitments that differs importantly from our own self-image.
- Because normative criticism challenges heartfelt norms, it is, as noted, powerless to *convince*. But if we *trust* our critics to the point of deeming their efforts *sincere*,[54] their implied account of our relevant commitments will still be deemed an *honest* portrayal, mak-

> ing us realize that others see us differently than we see ourselves. And the more trusted the critic, the more potentially disturbing the discrepancy.

Here, in the echo-chamber of keen yet trusted normative criticism, lies, we argue, the destabilizing or "ambivalating" potential of the intersubjective arena of normative discourse. When we consider our critics' efforts sincere, we find, refracted by that echo-chamber, depictions of our relevant commitments that differ worryingly from our own. Since the points of disagreement will necessarily pertain to the norms they question, the incongruity, as in the case of a disturbing playback, may have the effect of undermining our initial commitment to the point of rendering us *ambivalent* toward those norms. And, as Frankfurt famously argues, norms to which one becomes ambivalent lose their wholehearted volitional grounding, and with it their normative hold, and demand to be reassessed in the light of remaining commitments.

Wittgenstein speaks for many in asserting that, when "combating" people whose norms and standards differ radically from our own, there are only two possible outcomes: outright dismissal, when "each man declares the other a fool and heretic," or unreasoned "persuasion"—his term for "what happens when missionaries convert natives."[55] *The View from Within* presents a third option: to be rationally *motivated* by argument (though not *convinced* by it) to modify or replace our norms. But the level of self-estrangement and self-alienation it demands requires the potentially ambivalating intersubjective environment of trusted normative criticism, which far transcends the possible boundaries of even the keenest intrasubjective self-reflection.

The View from Within concludes with a short chapter entitled "Science Revisited," in which we briefly outline the possible relevance of our findings with respect to the role of agency, and the potential rationality of framework transitions in science. The present study is an attempt to make good on the promise of that outline.[56] With a better philosophical understanding of the dynamics of rational, deliberated normative change now at my disposal, the chapters that follow seek to amend the philosophical and historiographical approach of my earlier historical essays with a better suited alternative, arguing for it first in general and then by extended example, by revisiting my work on the history of early Victorian mathematics in the light of my newly acquired perspective.

PART 2

We (Philosophers)

CHAPTER TWO

The Philosophical Framework

THREE BIASES

From the purely autobiographical (and therefore thoroughly idiosyncratic) perspective outlined in chapter 1, the philosophical motivation for the present work originated in my awakening (prompted by my historical work) to the Popperian school's failure to appreciate the peculiar resilience of personal normative commitment to self-criticism, and the grave problem it poses to accounting for the rationality of significant normative realignment. I shall now take a more systematic and less personal view of the philosophical setting, field, and context relevant to the particular account of scientific agency this book seeks to explore.

But let me emphasize at the outset[1] that although this study focuses on the *rationality* of scientific framework transitions, it should by no means be read as an apologetic attempt to prove that science is, or is for the most part, rational in the strong sense outlined at the start. Its objectives are far more modest, and concern the *possibility in principle* of rationally holding our normative scientific commitments in normative check, characterizing the dynamics of such transitions when undertaken rationally, and deciding how their stories should be told. It is also important to realize that the opposite of rationality construed as being motivated by reasons is not irrational, but *a-rational* action. As in literature and the arts, new scientific and mathematical options are often explored playfully, and entertained for no better reason than to be original.[2] Not all mathematical innovations are motivated by pressing problems. Some are the outcomes of sheer playful curiosity—a desire to see what happens when this definition is reversed, or that axiom replaced by its negation. There is a categorical difference between realizing the value of a move in retrospect, and viewing it prospectively as the right move to make. This book is about the possibility of framework transitions of the latter kind.

Three latter-day philosophical biases or precommitments motivate the project to which the following pages are devoted—two of which hark back to major Kantian themes, the third, to what one might call their Hegelian corrective. All three have been alluded to, though not developed, in the preceding chapter. They combine to form the framework for what follows, in which they will be assumed and at points elaborated but will not be argued *for* except when diverting from accepted understanding or use. The present chapter discusses the first two. The third will be taken up in chapter 3.

FRAMEWORK DEPENDENCY

The first of these biases, central to the *Critique of Pure Reason*, is a firm commitment to the *framework dependency* of all judgment, with special reference to that of scientific knowledge. I shall leave the notion of "framework" deliberately vague at this point. In the work of some philosophers (such as Kuhn or Friedman), it is taken to denote a relatively well-defined subset of *knowledge claims*; for others (such as Sellars, McDowell, or Brandom), it signifies the interrelated *vocabulary* by which the body of knowledge in question is articulated; and for others still (such as Rorty), it indicates a subset of that vocabulary (a person's "final vocabulary," as he calls it). Kant's original notion of a set of forms and categories that jointly generate a body of synthetic a priori truths combines both approaches, of course, which renders these differences more a matter of emphasis than principle. The first group focuses more on the framework dependency of a judgment's *truth*; the second, on its *meaning* (or, more precisely, as McDowell and Brandom insist, its *intentionality*).

With respect to "dependency," however, I would like to be more precise. Framework Dependency is taken here in the strong sense of the second term: as denoting a relationship not merely of entailment or generation, but of *constitution*. Framework Dependency is the idea that a science's linguistic framework is what *enables* claims to scientific knowledge in its domain, without which such knowledge would not be *possible*. Linguistic frameworks are not what make such claims *true*, but what renders them claim*able*![3] They embody the standards on the basis of which judgments are issued, and with reference to which they are held accountable, and therefore mark the boundaries of the conceivable for those whose frameworks they are.

It is hence important to stress the extent to which Framework Dependency thus construed is a distinctly neo-Kantian notion, not to be confused with other construals of epistemic commitment. It premises a well-defined (if functionally rather than logically), two-tiered layering of scientific knowledge and knowhow,[4] one constitutive of the other. Quine, who has shown conclusively that the two cannot be logically differentiated, has famously argued for a form of epistemic holism that reduces normative commitment to mere entrenchment, to a "maxim of minimum mutilation" as he bluntly puts it.[5] Such a construal of commitment misses the entire point of the dependency relation, as Friedman, a staunch supporter of its two-tiered construal, forcefully argues in response to Quine.[6] Even if the two tiers cannot be distinguished formally, it is not the case that the distinction between them can be reduced to a continuum of mere relative entrenchment and relative resistance to revision, as Quine maintains, and the dependency relation obliterated. Friedman's wisely historical arguments pertain both to the specific function played by certain elements within a science (mathematical physics, in Friedman's case) at a given time, as well as to the ways in which the science as a whole was transformed over time. The way in which the calculus stands to the laws of motion in Newtonian mechanics, for example, cannot be plausibly described as being merely a matter of relative resilience to empirical disconfirmation.[7] The function of the calculus in Newtonian physics is constitutive. Without the calculus, Friedman argues, "[Newton's] second law of motion could not even be formulated or written down, let alone function to describe empirical phenomena,"[8] although the calculus was anything but well entrenched at the time! And the same goes for the mathematical theory of Reimannian manifolds and the principle of equivalence, which, while constitutive of Einstein's theory, were hardly well-entrenched parts of the physical or mathematical thinking of their day. It is during such revolutionary moments, argues Friedman, that the inappropriateness of Quine's attempt to reduce the a-priority of such elements to mere relative embeddedness becomes most apparent, despite his success in proving Carnap's attempt to demarcate them formally to be groundless!

Others, even within the Popperian camp, have gone important steps beyond Quine, but not quite far enough. The previous chapter made brief mention of Agassi's important observations regarding the role of metaphysics in raising and regulating scientific problems.[9] This is a far cry from mere entrenchment and relative resilience to empirical testing. Agassi accords metaphysical frameworks a major role in directing research and bringing certain problems to the fore by ascribing to them a deeper significance than others. According to Agassi, metaphysics is constitutive of *certain aspects* of the picture painted by science, by lending weight and substance to elements it regards as deep and basic. However, Agassi's contention that metaphysical frameworks can (and should) be appraised instrumentally against their interpretive efficacy and replaced when found wanting clearly indicates that the constitutive role he ascribes to them is severely limited in comparison to the notion of Framework Dependency adopted here.

Not unlike Agassi, Lakatos also acknowledges the presence of a body of "hard core" commitment at the heart of all scientific research programmes. Lakatos is less interested than Agassi in the role of a programme's hard core in generating and focusing scientific problems, and more interested in the central role commitment to them plays in the defensive tactics that adherents of such programmes employ in response to empirical problems. They are shielded from refutation by a protective belt of auxiliary hypotheses and an array of ad hoc moves that absorb the impact of empirical disproof and deflect it away from the hard core. If Agassi 1964 can be viewed as an attempt to extend Popper's philosophy to the realm of metaphysics, Lakatos 1970 shows how it is violated with regard to a programme's deep commitments. In this sense, Lakatos shows greater awareness than Agassi to the hold such framework commitments exert on the thought and practice of a program's practitioners. However, his account of the way *competing* research programmes are, or should be, adjudicated falls back on a form of instrumental reasoning similar to Agassi's, in which empirical fruitfulness trumps meta-scientific commitment.[10] From the neo-Kantian perspective to which the present study is committed, the existence of such an external, comparative point of normative reference is a myth.[11]

It is also important to emphasize, even at this early stage, that, as in Kant, Framework Dependency thus construed makes *two* claims, not one: one about the *produce*, and one about the *process* of science. A science's linguistic framework is constitutive not only of the world-picture produced by its practitioners, but of their very thinking and practice. The point is obvious, yet needs to be made. Is not the produce of science produced by its practitioners' practice, and therefore wholly determined by it? In which case, does it not follow that the constitutive dependency of the produce must be exclusively mediated by that of the practice? And in that case, does that not render the product's constitutive dependency on the framework *derivative* of that of the practice? As in Kant, the answer to all these questions is a firm "yes"! Framework Dependency is first and foremost a thesis about the constitutive force of a science's linguistic framework with regard to *agency*, the ability of the individuals whose framework it is to think, to reason, and to act. As we shall see, although from a neo-Kantian perspective it states the obvious, neo-Kantian studies of science tend to ignore this point and—for reasons to be discussed below—drive a wedge between the framework dependency of scientific product and process. They uphold the former, while (as in Friedman's case) making light of or (as in DiSalle's) ignoring the latter.

THE SELF-CRITICAL SELF

The second bias that grounds this study is an equally firm commitment to the essentially self-critical stance central to the kind of reflective, free-willed self-governance on which normativity and agency are made to turn in Kant's *Critique of Practical Reason, Foundations of the Metaphysics of Morals*, and *The Metaphysics of Morals*. In attempting to justify my ongoing commitment to Popper's identification of rationality with self-criticism, the previous chapter detailed the extent to which self-criticism thoroughly pervades rationality's two main intrasubjective stages: the *endorsement* of reasons, and their subsequent *application* in acting.[12] The second bias of which I speak mainly concerns the former process—that of responsibly determining what one deems right, and obligating oneself to be so motivated. As Kant has taught, the crucial distinction in this respect is that between being merely *subject to*, and *subjecting oneself to*, a law. It is the difference between being propelled by the causal impact of external force or internal impulse, and being *self*-propelled by freely submitting to a law of one's own making. The law, norm, or standard in question may, of course, have been thought up and legislated by others—they usually are. It is not required that we ourselves make them, but that we *make them ours* by a willed act of *endorsement* and self-submission. If one abides by *X* for fear of rebuke or punishment, or hope of reward, one is motivated by *X*, not by one's commitment to *X*. In this case one has not made *X* one's norm, and therefore one cannot be considered as responding *normatively* to *X*. The essential difference between being motivated by norm and being propelled by external force (from within or without) is—again according to Kant—that the former necessarily involves an uncoerced and autonomous exercise of free will. A forced or involuntary act, advantageous as it may be to one's most vital interests, cannot be deemed *normative* (or rational, in that it was not done or performed for a reason). The space of normativity is thus the space of agency—the space of reasons, the space in which a stand is taken and an act performed *because* it is freely and autonomously deemed the right thing to do.[13] Charles Taylor puts it concisely:

> To be a full human agent, to be a person or a self . . . is to exist in a space defined by distinctions of worth. A self is a being for whom certain questions of categoric value have arisen, and received at least partial answers. Perhaps they have been given authoritatively by the culture more than they have been elaborated in the deliberation of the person concerned, but they are his in the sense that they are incorporated into his self-understanding . . . My claim is that this . . . is essential to what we would understand and recognize as full human agency.[14]

Oppositions Right and Left

When taken on reflection, this idea of self-legislation and self-motivation is a position that is impossible to coherently deny. And yet, it is seriously contested both from the right and from the left. I shall not be arguing for it, but the opposition should nonetheless be at least noted. The very idea of a noncausal motivating force is seriously challenged from the right by the naturalistic sciences of man that a priori do not recognize active factors that are not causal. The various branches of neuroscience and many areas of cognitive science that follow suit, as well as much of decision and game theory, study human reasoning, cognition, and action by means of reductive explanatory repertoires—chemical, neural, computational (that is, algorithmic)—that are limited *in advance* to the bluntly causal. From their perspectives, acting in the space of reasons, responding to argument, being compelled to act by the force of commitment, by judging a course of action to be *right*—not to be merely in one's interest, but to be appropriate, fitting—is reduced by their explanatory strategies without remainder to being acted *upon* in the space of law or cause. The feeling of being moved by the normative force of reason is hailed as illusion, not because it is *found* to be illusionary, but because it is so assumed a priori.

Challenges from the left are more subtle and more difficult to meet, and more immediately relevant to the present project. Selfhood, in the sense to which this study is committed, is denied here also by reduction—not to brute physical forces, however, but to those of social conditioning. The self or subject is considered so wholly a *product* of social forces as to deny it any independent life of its own. The language with which one speaks and thinks, the social networks to which one belongs, the social norms, standards, standing, and station by and from which one appraises oneself and one's circumstance, are all *social* institutions that are deemed constitutive of the self without remainder. The many are held to determine the one, society to determine individuality, community to determine membership, so exclusively and exhaustively as to render the latter categories derivative to the point of redundancy with respect to the former.

As we shall see further down, the opposition from the left has gained real, if largely unwitting, ground in science studies, and not without reason—not because the collectivist position is deemed correct, so much as because the alternative position, which the present study seeks to promote, is deemed fatally problematic. With no intention to enter the fray in any detail at this point, I shall only say that while collectivist modes of explanation may be deemed relatively successful in accounting for situations of social stability, they have far less to offer to the understanding of social change. No one disputes the fact that the social and cultural institutions and structures by which individual agency is explained away in collectivist accounts are all nonetheless the products of human activity, and are subject to change, even radical change, again, as the result of human action. But if individual human reasoning and action is conditioned and reducible so thoroughly by and to these structures, how is their modification and replacement to be accounted for—especially in cases like scientific framework transitions, which we would like to think of as contrived, reasoned through, and quite deliberate. We shall return to this important point later.[15]

Normativity's Givens: Korsgaard, Taylor, Frankfurt, and McDowell, Briefly

Varied as their approaches are, those who do concede a nonreductive notion of human agency and offer serious philosophical accounts of normativity and personhood, in which individuals are portrayed as obligated rationally to taking responsibility for their reasoned choices, all preserve a firm element of normative self-criticism inherently and essentially at play at the heart of their accounts. And these accounts, I shall argue, all assume a given normative stopping point in order to ground the process. Without presuming to exhaust the field, let me briefly mention four such accounts—those of Christine Korsgaard, Charles Taylor, Harry Frankfurt, and John McDowell—all of which will bear in one way or another on the ensuing discussion. I shall then turn briefly to the intrasubjective chapter of Robert Brandom's similarly neo-Kantian account of selfhood and agency. It is very different from the other four and, in itself, I shall argue, quite unsatisfactory.

Brandom's importance for the current project lies in the second, decidedly intersubjective chapter of his account; its attempt to rectify the flaws of the first will pave the way to the third, "Hegelian" bias, or set of commitments, that underlies this study.

Korsgaard. The previous chapter briefly discussed Korsgaard's notion of "reflective endorsement," the critical enterprise of deciding which of one's impulses, sensations, and desires merit endorsement as reasons. Taking such a reflective stance requires one to self-divide and hold oneself in check. Harry Frankfurt, to whose account of personal identity we shall return shortly, describes it as a form of self-alienation, in which "a sort of division within our minds" is introduced that "establishes an inward directed monitoring oversight."[16] The self-consciousness achieved is more than mere awareness. The part assigned the "monitoring" is charged with normative responsibilities. It appraises and evaluates one's desires and impulses, judging some desirable and others adverse, some appropriate, others offensive. To assume agency is to undertake to resist the force of inner impulse and external constraint, and be motivated only by incentives one freely deems fitting. We do not always succeed. Even the most sincere normative aspirations are often defeated by superior forces, both within and without. But agency is a function not so much of normative triumph as of normative self-critique. It is in taking a normative stand against ourselves that we establish our agency and normativity.[17]

But on what normative grounds is this initial judgment supposed to rest? By what criteria, asks Korsgaard,[18] is one supposed to distinguish desires that merit endorsement as reasons from those that do not? Her answer builds on, but also interestingly veers away from, Kant. Like Kant, she maintains that for reflective acts to be at once autonomous and self-legislative they can be governed only by the idea of lawfulness itself (anything else would be deemed an external constraint). They are therefore subject to the categorical imperative: the demand that one's maxims be lawlike. But unlike Kant, she insists that universality comes in degrees. Applying a maxim universally to fellow males, or Muslims, or parents, or community members suffices to render it *normatively binding*. Only when applied maximally to the Kingdom of Ends, "the republic of all rational beings," does a maxim yield the moral law.[19]

One's reasons and obligations are selected, according to Korsgaard, on the basis of the particular "practical identity" one forms in accord with the "description under which [one] finds [one's] life to be worth living and [one's] actions to be worth undertaking."[20] A person's membership in the community, her citizenship, and her various roles as daughter, mother, or spouse, as adherent to a religion, as advocate of ideals and positions, as a professional of standing, and so forth combine to provide her with a not easily managed set of criteria for singling out the desires and impulses worthy of acting upon, and those fit to repress. One's practical identity is the ultimate source and grounding of one's normativity.

One's practical identity delineates the self one values, the self one aspires to be and would like one's actual, or empirical, self to live up to, which at times it does not. It is this self that, speaking for one in the first person, sits in perpetual judgment of one's actual self. Here lies the fierce, self-critical core of reflective endorsement in the form of a tensely, at times incongruently, divided or doubled self, forcefully captured by the "astonishing but familiar 'I couldn't live with myself if I did that!'"[21] For the sake of brevity I shall dub them the I-self and the A-self (*I* for "ideal," *A* for "actual").

The question Korsgaard leaves open, however, is that of the I-self's own normative grounding. She considers it a given. The I-self provides the normative criteria *for* reflective endorsement reasons, but is not itself described as reflectively endorsed. Korsgaard portrays people as undertaking commitments by means of normative vocabularies of valued self-description that *they already possess*. The I-self requires constant maintenance. Questions of normative priority and the need to adjudicate incompatible commitments arise all the time. Being an agent, Korsgaard implies, commits one first to ensuring the integrity of one's I-self, and then to critically applying it to one's actual self. What her account leaves open is the question of normatively *determining* one's I-self. Scrutinizing one's evaluative vocabulary for consistency and coherence is a matter quite different from holding it in *normative review*. This is the sense in which it is treated by Korsgaard as a *given*. Serving as the normative yardstick by which one's A-self is held in check, one's I-self is supposedly rendered normatively immeasurable.

Korsgaard's text strongly suggests, as do those of the three other thinkers I shall briefly discuss, that the ultimate source of our normativity is located in who we aspire to be, and who we aspire to be is at any one moment a contingent, given *fact* of our personal makeup. The exercise of agency is an exercise in normative self-criticism for which a normative framework is required, and for that very reason, it is more than implied, the normative framework cannot itself be considered the *outcome* of an exercise of agency.

The present study focuses on scientific, rather than ethical, normative outlooks. The very notion of a scientific paradigm nicely captures for scientific communities and their members the role played by practical identities in Korsgaard's account of personal identity and agency. Paradigms are practical identities in that they represent and embody internalized accounts of what communities hold worthy in their scientific work and produce. The comparison to science brings out even more clearly the extent to which the Korsgaardian model limits itself to what Kuhn called "normal science"—periods during which a stable paradigm provides the standard for the normative work of scientific training, research, and evaluation—but offers little to explain paradigm *shifts* as acts of rational agency. Korsgaard speaks for many in this regard.

Taylor. Charles Taylor's early studies of agency and personhood steer close to Korsgaard's picture. But they lay special stress on an aspect of agency that she mentions, but fails to develop: its articulative, linguistic nature.[22] Human agency, argues Taylor, uniquely asserts itself in acts of self-evaluation that go beyond merely choosing *between* the pull of incompatible first-order desires. He distinguishes between weak and strong self-evaluations. Weak evaluations assess and prioritize on the basis of outcome: Would one prefer vacationing in the North or in the South, dining here now or later at the beach, stopping at the red light or running the risk of being caught and fined? Were the options compatible, no question would have arisen. And their incompatibility is a matter of contingent fact. One would have liked to enjoy the recreational offerings of both the North and the South, but it so happens that one cannot. Strong evaluations, by contrast, assess first-order desires according to their *worth* by means of a vocabulary of thick normative concepts, such as brave, insulting, modest, considerate, and so forth—concepts that combine to depict the kind of person the evaluator aspires to be.[23] Giving in to a desire that runs afoul of one of a person's normative concepts—succumbing to the impulse to flee, for instance, by someone who values "a courageous and upright mode of life"—is considered not a missed opportunity, but an act of *self-betrayal*![24]

Such evaluative concepts, Taylor notes, are by nature contrastive. To rule normatively in favor of certain modes of conduct is, willy-nilly, to rule *normatively* against others. This is because strong evaluation is not an act of choosing between various appealing options, or of stating comparative preference in the face of contingent incompatibility, but an act of intrinsic *approval* of certain forms of conduct to the *condemnation* of others. Only in the latter case can there be a discrepancy between what one desires and what one deems worthy. Deciding (freely) merely according to preference—between vacationing in the North or South—can, by definition, never run up *against* what one mostly desires. Weak evaluation is certainly a critical enterprise, but it is not a *self*-critical one. Strong evaluation is. Ruling normatively can certainly run squarely up against one's heartfelt desires and is, hence, inherently self-critical. For it asserts itself between what one desires and what one *ought* to do or to think.

The point Taylor makes central is the extent to which strong evaluation is an act of *conceptualization*. Humans are perhaps the only species capable of *deliberating* how best to prioritize and decide between conflicting desires and impulses, and between different means of satisfying them, but they are by no means the only species that faces such dilemmas. Strong evaluation , by contrast, is an exclusive, as well as a defining, human undertaking. It is a process not merely enhanced by our linguistic capabilities, but wholly *dependent* on them. Normativity, the ability to possess and apply standards of *propriety* and *significance*, requires, or rather *is*, a particular act of conceptualization mediated by an essentially anthropocentric language of contrastive evaluative distinctions,[25] such as "noble or base, integrating or fragmenting, courageous or cowardly, clairvoyant or blind, and so on."[26] We deploy our moral vocabularies in *discerning* who we are and who we want to be, and in shaping our actual selves accordingly. They thus play an ongoing, yet ordered double role: the first, *self-interpretive* or, as Taylor prefers, *self-expressive*; the second, normatively *self-critical*. To understand what it is to be a person is to understand the intrasubjective, first-person normative, self-descriptive, and self-corrective discourse of strong evaluation—a task at which naturalist, objective, third-person approaches to the study of man, Taylor firmly argues, all miserably fail.[27] Our normative vocabularies are born and refined in an ongoing effort to describe not who we *are*, but who we *aspire to be*; that explains their powerful *prescriptive* force with respect to the former, which Taylor accounts for in decidedly romantic terms, as a dialectical process of mutual *informing* that vacillates back and forth between *interpretandum* and *interpretance*.[28]

Taylor is an interpretivist in this regard, which enables him to account, better than Korsgaard does, for the way our normative vocabularies change over time. But similarly to Korsgaard, for him, too, the exercise of human agency in strong evaluation requires an *already existing* normative vocabulary, our commitment to which cannot itself be attributed to the kind of strong evaluation of which it is constitutive—which, again, is to limit the exercise of agency to periods analogous to "normal science."[29]

Although Taylor nowhere states the problem in these terms, I believe it was felt and dealt with in his later work by introducing the notion of hypergoods[30]—goods, like the search for justice and the love of God (Taylor's own two examples),[31] located incomparably higher than any other in our system of values, that by virtue of their incomparable importance provide the normatively critical standard by which our other goods must be weighed and judged. Hypergoods thus provide the basis for normatively criticizing our "practical identities" as a whole—the element missing in his earlier work.

But this is not to solve the problem, only to shift it to the higher level of hypergoods. Can commitment to the hypergood itself be subject to strong evaluation? Taylor argues that it can. It is a somewhat odd argument. "Is there any rational way," he asks, "for *A* to convince *B* that his hypergood perspective is superior? And if not, then how does *A* rationally convince himself?"[32] To conflate the two questions, Taylor argues, is to misunderstand the very nature of the domain of moral reasoning, by wrongly subjecting it to the naturalist criteria of scientific deliberation. Arguing in favor of a scientific theory requires bracketing one's "anthropocentric" perspective and offering arguments that are neutral across such divides. In the thoroughly anthropocentric realm of the moral, such a stance is unavailable. It is impossible to convince a person who shares none of our basic moral intuitions of the justice of our cause, but that does *not* mean that we cannot convince ourselves. We can do so, he argues, from a *genealogical* perspective. "A hypergood," he writes, "can only be defended through a certain reading of its genesis"[33]—by seeing, from the perspective of the moral intuitions we possess, and in the light of what we are moved by morally, how, in the course of its development, our present view of justice, for instance, gained in scope and generality by application first to religious minorities, then to slaves, then to women, and so on. Viewing it as the present culmination point of a nested process of expansive articulation, he submits, provides us with a perspective both to justify where we stand and to anticipate further development.

Taylor's position raises two related problems. He is right to contrast the kind of expressive, first-person normative evaluation associated with human agency with the seventeenth-century ideal of objectified third-person scientific inquiry. However, he writes throughout as if the distinction remains philosophically sound today—as if one can apply the romantic picture of human reasoning and judgment as thoroughly dependent on a framework of internalized normative concepts only to the moral sphere, to the exclusion of that of science. This amounts to incoherently acknowledging the normative basis of agency and personhood in general, while denying it to science, as if the norms of scientific conduct and assessment are essentially different from other norms. But Taylor's philosophy of science is not our concern here.[34]

The second problem Taylor's notion of hypergoods raises is more significant. The idea that sufficiently diverse normative outlooks are normatively unrankable is one I concede and have argued for elsewhere.[35] I also concede Taylor's conclusion that therefore a normative outlook can only be normatively appraised from within. I also see the point of claiming that the normative genealogy of an outlook can provide noncircular (or not fatally circular) grounds for *approving* its current culmination point. But strong evaluation has less to do with normative *affirmation* than with normative *criticism*. Even if awareness of a hypergood's genealogy is capable of deepening its adherents' commitment to it, it is quite incapable of providing them with grounds for its normative *critique*. This is because stages in the development of a normative outlook may seem nested retrospectively, but are never really so. Michael Friedman's account of the rationality of framework transitions in science also lists the nestedness of successive paradigms as a reason for practitioners of the old to adopt the new. But this is an illusion. Although Einstein's special relativity preserves Newtonian physics as a special case, the reality it prescribes and the "space of intellectual and empirical possibilities" to which it gives rise are radically different from those premised by Newtonian physics. Moving from the former paradigm to the latter required finding the former normatively *wanting*, not merely unduly limited in *scope*. The development of physics until that point in time has nothing to offer in that respect. Thus, despite Taylor's good efforts to the contrary, Taylorian hypergoods are no less a given than Korsgaardian practical identities. Still, although he fails to solve it, Taylor at least senses the problem.

Frankfurt. Both Korsgaard and Taylor rightly acknowledge their debt to Frankfurt's seminal 1971 essay "Freedom of the Will and the Concept of a Person"[36]—Korsgaard notes its "affinity" to her views,[37] and Taylor, as we have seen, takes it as his point of departure[38]—but neither of them properly engages Frankfurt's later work, especially that related to caring and love.[39] They explain how one's normative vocabularies are marshaled in reflective A-self governing, but, unlike Frankfurt, do not explain the nature of their hold and claim to one's commitments. Distinguishing between imagined I-self (practical identity, the vocabulary of strong evaluation, hypergood) and actual A-self (first-order desires and impulses) rightly assumes, but of itself does not explain, one's dedication to the former. Indeed, the metaphor of a divided or split self can be misleading in this respect. For it is true that the I-self is an imagined self, and as such it is distinct from its actual or empirical counterpart, but to command the assent it does, it must also be deeply *desired* and hence firmly rooted in the actual self.

The I-self presents us with much more than a feasible *option*. Its *potentiality* resides in the specific imagined vision of selfhood on which to set our sights. But the authority with which it speaks for us and its power to mobilize us to act on its behalf bear clear witness to a robust *actuality* grounded in who we already *are*. And it is this actuality that, according to Frankfurt, renders our I-self *ours* and explains our commitment to it. This is the key to his hierarchical and twofold account of personal identity. The I-self is the self we are not, but *want* to be, and for it to be not only imagined, but desired, it must spring from our volitional makeup.

Like Korsgaard and Taylor, Frankfurt locates a clear and firm element of normative self-reckoning and self-criticism at the heart of the basic workings of human agency. Again like them, he presupposes as necessary an extracritical, pregiven normative grounding, but unlike them, he undertakes to explain this. Discerning who we really are and deliberating who we ought to be jointly define the Janus-faced project of "Taking Ourselves Seriously" and "Getting It Right," to quote the titles of Frankfurt's two Tanner Lectures that form the basis for Frankfurt 2006. First, the shape of the will and the limits of our autonomy are determined by taking *volitional stock* of our feelings and desires, and our dispositions to act in certain ways. Then, "we disrupt ourselves from an uncritical immersion in our current primary experience, take a look at what is going on in it, and arrive at some resolution concerning what we think about it or how it makes us feel."[40]

Such "resolutions" amount to selectively identifying *with* certain of our desires and dispositions, and making them our own, while rejecting and externalizing those with which we refuse to identify. To identify with a desire or impulse is to "recognize them as grounds for deciding what to think or what to do"; to refuse to do so is, by contrast, to "deny them any entitlement to supply us with motives or with reasons."[41] This twofold process of identification, first in one sense of the term, then in the other, is the way we constitute ourselves as agents—reflective and normatively self-critical.[42] But the normative adjudicative process of thus forming one's personal identity requires that a higher-order set of "attitudes and responses" be in place. Hence Frankfurt's basic distinction between first-order and higher-order volitions.

What lends the I-self its special judicial status, he argues, is its volitional grounding, not merely in what we want in earnest, but in what we earnestly *want to want*. Not all philosophers have fully appreciated Frankfurt's notion of second-order volition. Taylor, for instance, thinks it is defined too broadly to distinguish between "strong" and "weak evaluation"—as if Frankfurt would consider deciding between having fruit salad or crème brûlée for dessert a matter of second-order volition. Taylor is wrong. Assessing which of two first-order desires is the most easily or conveniently satisfied, or which of them promises the most satisfaction, and so forth, does indeed involve a second-order evaluation of first-order desires, but this has nothing to do with Frankfurt's distinction. To assess the pros and cons of two mutually incompatible, first-order desires, with a view to discerning the stronger pull, the more satisfying outcome, or the less risky or costly of the two, is to assess which *outcome*, not which *desire*, is the more desirable. Second-order volitions in Frankfurt's sense are not employed in comparing and contrasting the strength or immediacy of first-order desires.[43]

A second-order desire, in Frankfurt's sense of the term, is a desire to have or not to have certain *desires*. Second-order volitions take an approving or disapproving stand with regard to first-order *cravings* deeming them *wanted* or *unwanted*. Becoming aware of a desire to throttle a critic or a craving to take a second helping of dessert and desiring not to have them, or desiring to have a desire one doesn't have, or desiring to have a stronger desire than one has to exercise, diet, or study more studiously is to pass judgment on oneself, on one's very volitional makeup. Such evaluations are, therefore, by definition, always normative and always strong in Taylor's sense of the term. One's second-order volitions thus circumscribe and define the person one wants to be, the self one wants to have. Jointly constituting the I-self, they sit in perpetual judgment of the rush and gush of one's first-order desires, incentives, and impulses, pronouncing them desirable or undesirable, worthy or unworthy, with an authority to self-rule that derives from the deepest reaches of one's will.

The stamp of approval issued by a second-order volition is that of *wholeheartedness*—one of Frankfurt's most important words. Wholeheartedness is a measure, not of the *strength* of a desire, but of its subjective *rightness*. To want something wholeheartedly is to deem *wanting it* desirable. Second-order volitional self-rulings hence derive their authority from being autonomous in the deepest sense of the word. For what is free will, asks Frankfurt, if not wanting wholeheartedly what one *wants* to want?

The particular form of yearning for *X* expressed in wanting to want *X* and the particular form of dedication to *X* to which it gives rise mark *X* as something or someone for which we *care*, for we do not merely want it, but find our very wanting it meaningfully pleasing. Desiring the desire for *X* marks *X* as a source of *self*-approval. *X* is valued for more than the satisfaction it gives us. Because we want to desire it, our desire for *X* carries a distinct measure of *self*-value. Therefore, failing *X* becomes a form of self-betrayal. This is for Frankfurt the source of what one deems to be right, and of one's commitment to pursuing it. The source of normativity is located, not in the law or the principle by which the free will operates, as in Korsgaard's Kantian formulations,[44] but in *what* the will most wholeheartedly wills.

For Korsgaard, the source of our commitment to the self-legislations of the I-self lies in the formal idea of law itself—in the way our maxim is rendered imperative when freely made a law. For Frankfurt, the normative and our commitment to it are grounded in the particular *contents* of our higher-order volitions. The people, the things, the forms of conduct, the ideas and ideologies that one loves, cares for, and deems important are those one wholeheartedly wants to want, and hence they are those that give rise to one's standards of propriety and obligation to act and abide by them. This has little to do with the act of lawmaking per se. The nature of the imperative is different from Kant's and Korsgaard's, and is grounded, according to Frankfurt, in the "volitional necessities" that "guide and limit [a person's] agency" by determining "what he may be willing to do, what he cannot help doing, and what he cannot bring himself to do."[45] Frankfurt distinguishes between two related categories of volitional necessity: the necessary constraints of the *unwillable*, and the necessary entailments of love: what we are incapable of willing ourselves to do, and what we *cannot but* will ourselves to do. The former entail the strictures of *volitional rationality*. Being volitionally rational "involves being incapable, under any circumstances, of making certain choices"—not because we lack the capacity or skill, but because we are incapable of mobilizing the will to do so, because they are unthinkable for us.[46] The dictates of wholehearted love, in contrast, define the normative positively. They determine what we should care about, who we cannot help being and wholeheartedly wanting to be. Necessities of this kind are twofold:

> First, a person cannot help loving what he loves; and second, he therefore cannot help taking the expectation that an action would benefit his beloved as a powerful and often decisively preemptive reason for performing that action. Through loving, then, we acquire final ends to which we cannot help being bound; and by virtue of having those ends we acquire reasons for acting that we cannot help but regard as particularly compelling.[47]

This is how the higher-order volitions that mark which first-order desires one wholeheartedly wants to be one's will[48] define and ground both the particular nature and structure of one's I-self and its normative authority over one's thought and action. To function properly as full-fledged agents "requires a clear self-understanding and a wholehearted acceptance of the essential requirements and boundaries of our will. This amounts to finding a mature confidence . . . in which the authority of our norms of conduct are grounded . . . a confidence in what we cannot help being."[49]

However—and this is the important point in the present context—even more markedly than Korsgaard's practical identities and Taylorian hypergoods, the Frankfurtian volitional necessities that give rise to the procedures, standards, and decisive rational warrants that supposedly decide what one *should* care about are determined exclusively by what one *does* care about already. They provide powerful normative strictures, but their bootstrapped grounding (to use Barbara Herman's term)[50] would seem here also to render them immune to normative appraisal. Frankfurt's response to Herman, whose account of his views he warmly deems "extraordinarily comprehensive and accurate," says it all:

> In my view, there really is no authority for us other than the authority of what we care about—or, more particularly, the authority of love. What we love is what we care about in a way that is not up to us. It is its necessity, together with the fact that it defines our will, that provides it with authority . . . This has very little to do with reasons or with a response to value.[51]

Frankfurt's picture of personal identity thus resembles those of Korsgaard and Taylor in depicting agency as constituted by the normatively critical tension between I-self and A-self—in his case, between second- and first-order volitions—in understanding the former as the seat of our norms and standards, and hence as a necessary condition for the exercise of agency, and in insisting, therefore—in his case explicitly—on their givenness and immunity to reasoned normative evaluation. In Frankfurt's picture this last claim receives greater depth and urgency than in those of Korsgaard and Taylor, in the way it grounds the normative given in reaches of the will located so deep as to no longer be under our control, as he admits explicitly.[52] John McDowell, to whom we now turn, locates what we are wholeheartedly committed to "in a way that is not up to us," in second nature.

McDowell. Unlike the three thinkers so far discussed, McDowell has shown little interest in questions of personhood, identity, or agency as such. Some of his early writing deals with the role played by belief in moral reasoning, the nature of moral reasons, and their authority (ideally) to overrule and supersede other reasons to act.[53] Like the other three thinkers, he concedes the idea that normative judgment, exercised intrasubjectively to override other considerations, or intersubjectively in criticism of others, requires a prior grasp of normative concepts and an array of pregiven commitments. Unlike the other three, however, McDowell's central concern, at least since *Mind and World*, is not that of applying one's normative thinking to putting oneself right, nor even, as in Frankfurt 2006, that of getting oneself *factually* right.[54] The problem he addresses, as noted briefly in the previous chapter, is that of getting one's normative thinking *normatively* right—which, for him, means holding it answerable to normative criticism from within that thinking.[55]

But, as we have seen, McDowell does little toward showing how assuming such a stance in ethics is at all possible. His allusion to Neurath's famous image, and talk of ethical outlooks standing up to *their own* reflective scrutiny, strongly suggests that what he has in mind is troubleshooting for incoherence, lack of clarity, or inconsistency, rather than for actual ethical failing. In any case, the point is developed no further with regard to ethics in *Mind and World*, or in any of McDowell's subsequent writings. But ethics has not been his main concern. The lion's share of McDowell's work since *Mind and World* is devoted to the question of getting our *empirical* thinking right[56]—a discussion that would seem more relevant to the concerns of the present study than his remarks on ethics. Here he no longer remains merely declarative.

The problem of the normative accountability of a worldview or framework is in general that of what Alex Byrne nicely dubs "spin control."[57] By what standard can our very standards of right and wrong be deemed wanting? What do they rub up against that can be said to keep them in normative check? And if they do not, and spin freely, to what can they be considered rationally relevant? Indeed, what can they be said to be *about*? McDowell addresses these questions primarily to the framework by which conceptual content is imparted in judgment. Promisingly, he formulates the question as a concern about rational framework *amendment*. How can our worldview be rationally adjustable in response to our experience if it is constitutive of our experience?[58] A "bald naturalist," such as Quine, untenably conceives the preconceptualized impressions the world makes on us as the tribunal of our empirical thinking, "even while he understands experience in terms of irritations of sensory nerve-endings."[59] At the other extreme, equally untenable for McDowell, philosophers like Davidson deny that such purely natural happenings can impart any content, and therefore, denying empiricism altogether, advocate a thorough coherentism that views our connected system of concepts and conceptions as "spinning frictionlessly in the void" of a space of reasons of its own device.[60] The middle Kantian ground McDowell purports to locate between the two extremes is a view of perceptual judgment held accountable to our spontaneous, yet contentful, perceptual experience—or as he has it: to the involuntary "actualization, together, of conceptual capacities whose active exercise, with the same togetherness, would be the making of a judgment."[61] McDowell usefully locates the source of these conceptual capacities in "being initiated into language,"[62] and their specific contents, by virtue of that initiation, as belonging to our second nature.[63]

The "spin control" McDowell ultimately offers is answerability not to a more fundamental set of experiential norms and standards, but to the world we spontaneously experience by virtue of second nature—the internalized conceptual contents we actualize involuntarily in perceiving. This is a form of interpretivism, and it is therefore unclear how it purports to transcend pure coherentism. For perceptual experience to constitute a meaningful tribunal, it is not enough that it be of sufficient conceptual content to merit placement in the space of reasons. To provide a normatively meaningful tribunal, our perceptual experience must minimally incorporate conceptual content that is not of our own device. As McDowell has it, empirical accountability boils down to subjecting our *considered* judgments to the tribunal of those we make unthinkingly. If any real critical space is opened by McDowell's very limited notion of empirical accountability, it is between what we take our concepts to be, and those we tacitly employ in perceptual experience. But favoring the latter would seem to rob the former of any normative force.[64]

But then, normatively privileging the latter is also unclear. In scientific research, for instance, it is the less-considered latter that, as a rule, is made answerable to the better-considered former, rather than the other way round. McDowell does not write about science, and does not address the problem of the rationality of framework transition in or outside science.[65] His interest lies in "controlling the spin" of a given framework with respect to the world it purports to describe, by presupposing discrepancies between perceptual experience that owes directly to the constitutive force of the normative framework, and the less-educated perceptual experience that owes to other forms of second nature.[66] Translated into the terms we have been using, the studied, cultivated conceptual vocabulary responsible for considered experiential judgments corresponds, we might say, to one's I-self, and that exercised in unstudied, involuntary, spontaneous experience, to one's A-self. Viewed thus, the normative force of the former again becomes unclear. To what, exactly, is one committing oneself if A-self is allowed to trump I-self and hold it in normative check? Unless, of course, second nature is granted normative priority across the board, in which case, our studied, cultivated frameworks cease to resemble I-selves altogether, and are rendered hypothetical attempts to articulate and explicate the conceptual contents of second-nature perception, to which they are forever accountable.

I doubt such a position could be fairly ascribed to McDowell. There is an obvious interpretivist whiff to his notion of second-nature accountability, but with regard to the constitutive role of conceptual frameworks in considered judgment, he is quite clear, yet far less clear as to how to coherently reconcile the two.[67]

McDowell's account, like those of the other thinkers we have considered, contains a solid and central element of normative self-criticism, which by going in the opposite direction from them (from A-self to I-self, rather than the other way round), would seem at first blush to break the hold of the givenness of the normative ground for such criticism that characterizes their accounts. But it does not. McDowell's interest in empirical frameworks seeks, as in his early work in ethics, to safeguard a robust and rational notion of framework-dependent *realism*, which amounts, in the scientific case, to calibrating the conscious and tacit components of an up-and-running scientific framework. He, therefore, closely resembles the other three thinkers in confining his discussion to what Kuhn termed normal, rather than revolutionary, science, and leaves wholly unattended the kind of normative friction justifiably responsible for rational framework *transitions*. In this respect, his notion of second nature preserves a normative givenness (if admittedly framework dependent) akin to that of Korsgaardian practical identities, Taylorian hypergoods, and Frankfurtian volitional necessities. Locating the normative given in second nature lends it a firmer embeddedness and greater resilience to rational dislodging than even Frankfurt's "reasons of love."[68]

Brandom's Kant in Contrast

I now turn briefly to the first of the two components of Brandom's most recent work on normativity that he describes as Kantian and Hegelian, respectively.[69] In truth, as Kantian as the first chapter of his account purports to be, it diverges so much from the other four thinkers we have considered as to render it largely irrelevant to the present work. Still, it deserves brief mention for two reasons. First, for the stress it places—more than any of the other four—on one's obligation to what we might call *rational* I-self

self-management, which it singles out as the definitive feature of human selfhood. Second, and more important, for how it paves the way for the Hegelian corrective presented by the subsequent part of Brandom's picture, which, unlike the first, will prove to be centrally relevant to what I have dubbed the third bias that motivates this study.

Brandom introduces the Kantian chapter of his account promisingly enough by attributing to Kant "the idea that what distinguishes judging and intentional doing from the activities of non-sapient creatures . . . is that they are things that knowers and agents are in a distinctive way *responsible* for. Judgments and actions make knowers and agents liable to characteristic kinds of *normative* assessment."[70] The chapter is devoted to discerning the precise nature of agency's normative import. What is it, he asks in characteristically pragmatist terms, that "one must *do* in order to be taking responsibility for or committing oneself to a judgeable content (or practical maxim)[71] in the sense that matters for apperceptive (sapient) awareness"? According to the four other thinkers we have considered, one's duty in this respect is to subject candidate judgeable contents and practical maxims to the normative appraisal of one's core commitments—to apply to them the standards of *propriety* grounded in one's practical identity, vocabulary of strong evaluation, wholehearted second-order volitions, or second nature. For them all, taking such normative responsibility is to *apply one's norms* in keen second-order normative critique and judgment. Disappointingly, Brandom's notion of intrasubjective normative responsibility does not involve *normative* assessment at all.

True to the inferentialism that motivates his entire approach, Brandom defines the responsibility one undertakes in judging as an undertaking to *integrate* each new judgment one endorses into the bulk of one's previous endorsements. Synthesizing the body of one's commitments, he writes, echoing Kant,[72] "produces, sustains, and develops a synthetic unity of apperception"—which, going a step beyond Kant, he actually identifies with the rational "self or subject."[73] The integrated sum of one's commitments is hence the Brandomian equivalent of what I have dubbed one's I-self. As in the other four accounts, the I-self charges us with responsibilities, which Brandom, unlike the other four, identifies as wholly self-reflective. They pertain, he insists, not to the formation or the application of commitments, but to the coherence of their mutual integration—an undertaking that charges one with responsibility for three sorts of activity: *critical*, *ampliative*, and *justificatory*.

For Frankfurt, getting oneself "right" is to determine what one *really* cares for in the face of deluding tendencies to confuse first-order impulse with second-order approval. Brandom takes one's first duties as a human agent to be *rational* rather than virtuous—to be committed *for a reason*. The conditions and criteria for endorsing a judgment or a practical maxim as a reason and for a reason are, for Brandom, not normative, but strictly *inferential*. One's commitments must be *made explicit*—first by simple articulation, so that incompatible commitments can be located and weeded out (one's *critical* responsibility). Then, the inferential consequences of the commitments one retains must be traced downstream, to render explicit the commitments each of them gives one reason to accept (one's *ampliative* duty). Finally, to be able to offer reasons for the commitments one acknowledges, the *prior* commitments located upstream from them that "inferentially entitle" one to them must also be explicated (one's *justificatory* obligation).

Brandom's emphasis on a person's duty to integrate and develop his or her acknowledged commitments into a *consistent*, *complete*, and *warranted* whole is, in itself, a welcome supplement to the discussion so far.[74] The problem with Brandom is not his holism, but his notion of intrasubjective normative critique or, rather, lack of it. Taylor, we have seen, also gives pride of place to articulation in the definition and development of selfhood, but limits it to making explicit the normative vocabulary employed in *self-adjudication*. Brandom casts his net wider. For him, human agency does not assert itself primarily in one's capacity to keep *oneself* in normative check. Acts of responsible reflective *self*-judgment are for him not a matter of regulating and controlling one's desires and impulses in the light of one's higher aspirations. Rather, it is first and foremost a matter of formally squaring each new inclination to pass judgment on anyone and anything with the body of one's existing commitments. Perhaps because he does not see it primarily in a *self-governing* role, Brandom's conception of the I-self as a unity of apperception stresses its *unity* at the expense of the sort of apperceptive *normative hierarchy* that, as a matter of principle, Korsgaard, Taylor, Frankfurt, and even McDowell locate definitively at the heart of theirs.

Although the project of the first, intrasubjective chapter of Brandom's account harks back to Kant's First rather than Second Critique, it contains nothing that resembles the a priori fixed Kantian core or framework of higher-order intrasubjective normative preferences. Brandom's unities of apperception privilege language as a whole, but do not privilege a body of core framework commitment akin to those constituted by Korsgaard's practical identities, Taylor's strong evaluations, Frankfurt's second-order volitions, or McDowell's second nature. As a result, his apperceptive holism leaves entirely open the question of which of two incompatible commitments to relinquish and which to retain. Whether it is the new that is made to override the old or vice versa is of no normative consequence, as long as consistency is preserved. In other words, in proceeding consistently, one has fully discharged one's *critical* responsibilities as a rational and normative human agent.

Because normative immunity is not granted a priori to any commitment, Brandom's account of the self would seem to be far more dynamic and open to criticism than the other four, but its flexibility comes at the unacceptable price of lacking any clear notion of *normative* self-criticism. Put differently, the problem with Brandom's picture is its implicit denial of Framework Dependency at the level of intrasubjective normative deliberation, at least as understood by the other four thinkers. The framework on which Brandomian intrasubjective deliberation depends is not a privileged set of deep and fundamental value judgments, but the materially connected, inferential whole of one's entire language, for which the only normative obligation is to apply consistently. Brandom's picture purports to compensate for much of this by means of its second, Hegelian chapter, which highlights "the crucial role played by the attitudes of *others* . . . in constituting the authority [one] has to make [oneself] responsible to." Hegel's "reciprocal recognition model of normativity" is introduced as "a way of developing Kant's autonomy model, so as to make sufficient room for the relative *independence* of conceptual contents from the attitudes of those who endorse them."[75] This, however, is to conflate the source of normative *content*, and hence of normative criticism, with that of normative *commitment*. As Brandom notes correctly, "one is genuinely *normatively* bound only by rules one has bound *oneself* by,"[76] but to be normatively *bound*, the rule itself cannot be entirely up to one. Hence, the possible role of others in normative criticism, to be discussed further down. But to create sufficient "room" between intrasubjective normative attitude and intersubjective normative status, in order for the critique of others to take hold (which we shall see is notoriously difficult to account for), a robust account of normative *self*-commitment, *self*-criticism, and *self*-regulation is necessary. And that, as Korsgaard, Taylor, Frankfurt, and even McDowell well understood, presupposes a firm and given normative grounding that far transcends even the firmest Brandomian commitment merely not to contradict oneself!

BIASES COMBINED

As we have seen, the relatively stable, normatively privileged, core set of commitments, definitive of personal identity and constitutive of the normatively critical processes of self-appraisal that each of the four thinkers locates at the heart of agency, comes at the price of its own exemption from, or immunity to, normative scrutiny and reasoning. Instead of being the prize products of reasoned reflective endorsement or strong evaluation, our deepest and most definitive commitments seem necessarily required to be firmly and contingently in place for the commencement of such processes to even be possible. Because the hypergoods and volitional necessities form the basis for one's normative reasoning, they cannot be *given* reasons. And since they are not the *product* of normative reasoning and, therefore, do not derive from more basic commitments, one is incapable of thinking or acting against them. As Frankfurt puts it, a person "cannot organize himself volitionally in the necessary way. In attempting to do so, he runs up against the limits of his will. . . . For him willing to perform it is unthinkable."[77]

In other words, as I briefly noted above in discussing Brandom, commitment to the second bias necessarily entails commitment to the first. Commitment to the view that normative self-criticism is inherent to selfhood and agency entails commitment to Framework Dependency in the strong sense described at the opening of this chapter. In fact, it entails a broader notion of Framework Dependency than the one outlined at the outset, rendering framework dependent not only all judgment, but all meaningful, reasoned *action*. But with it comes the sobering and worrying realization that animates the present study—that taken together, the two biases firmly commit one to the idea that the normative frameworks that make reflective endorsement possible cannot themselves be endorsed or rejected reflectively.

CHAPTER THREE

The Problem for Science

THE POST-KANTIAN PREDICAMENT

In Kant's original works the two sets of presuppositions that philosophically inform this study—the first two of its three biases, as I called them—remain wholly and understandably unrelated. For Kant, the framework articulated and analyzed in the *Critique of Pure Reason* cannot be considered a *normative* system. The two forms of intuition, the categories, and the framework of synthetic a priori truths to which they give rise govern human cognition, in principle, as a matter of *fact*, not because they are in any sense of the term found *fitting* or *good*. They are conceived by Kant as fixed and universal facts of human nature that are disclosable by the sort of reverse cognitive and logical engineering he called transcendental deduction. They are facts on which we can reflect and gain progressive understanding, but concerning which *we have no say*. From a Kantian perspective, we cannot choose, or even seriously contemplate, seeing things differently; objecting, for example, to experiencing physical objects in Euclidean space and time or as impacting each other in anything that is not a causal relation. There is no coherent, strictly Kantian sense in which such an objection could even be articulated, let alone acted upon by seeking alternative ways of doing so.

But in the neo-Kantian picture of science that owes to the work of Moritz Schlick, Hans Reichenbach, Rudolph Carnap, and Kuhn, and has been developed most fully to date by Michael Friedman, this is no longer the case.[1] By relativizing the constitutive a priori—mounting "Kant on Wheels," as the late Peter Lipton so nicely put it[2]—scientific frameworks are no longer seen as belonging to human *nature*, but as acquired, self-fashioned components of human *culture*, internalized through scientific training. They are thus rendered *normative* systems by definition—systems that, in addition to determining for their practitioners what is valid, right, and worthy in science, are *deemed themselves* to be valid, right, and worthy; systems that rightly establish the standards of scientific procedure and propriety in the light of which new practitioners are trained and which they are expected to abide by, yet are at liberty, in principle, to obey or violate—and, hence, we would like to think, also capable of calling into question.

One would therefore have expected latter-day philosophers of science, especially those of neo-Kantian leanings, not to follow Kant in this respect, and to run the two principles that ground his First and Second Critiques centrally together. But this has not at all been the case. If in Kant they are relegated to separate Critiques, in latter-day philosophy of science they have been relegated to diametrically opposed schools of thought![3] Neo-Kantians, like Kuhn and later Friedman, who fully concede Framework Dependency, firmly resist the idea that, in order to be deemed rational, a framework's normative hold must derive from its passing normative critical muster, and, more important, that its replacement, if rational, must be motivated and justified by its apparent failure to do so. Popper and his school, on the other hand, while fully conceding the fundamental centrality of self-criticism to the rationality of all acts of scientific endorsement and rejection, largely ignore, and in most cases explicitly reject, what Popper dismissively describes as "the myth of the framework."[4]

For those who equate rationality with self-criticism (along the lines argued for in chapter 1, for instance), contesting "the myth of the framework," some might argue, would imply conceding some form of what Sellars famously dubbed "the myth of the given"[5] to provide critical footing. In Popper's case, however, this is not true—at least at the declarative level—since he insists on the "theory-ladenness" of even the most elementary perceptual experience. Rejecting both "myths," however, amounts to accepting, on the one hand, the idea that all general and particular assertions are conceptually informed, while denying, on the other, the existence of a conceptual framework constitutive a priori of what we can claim to know. To jointly maintain both these positions, and to insist in addition, as Popper does, on a (regulative) notion of truth as correspondence, and of scientific trial and error as exhibiting real verisimilitude, requires an account of the conceptual that would seem not to fall under any viable current rubric. Popper and his followers have failed so far to articulate such a theory.

The problem of articulating a coherent Popperianism in this respect is further accentuated by the conclusions of the previous chapter, where it emerged that identifying rational agency with adopting a *normatively* self-critical stance, as in the works of Korsgaard, Taylor, Frankfurt, and McDowell, *commits* one to the kind of Framework Dependency Popper adamantly rejects. (Even if self-criticism is limited to troubleshooting for material incompatibility, as in Brandom's account, Framework Dependency still necessarily obtains, but is extended to the connected, materially inferential web of one's particular language in its entirety.) Only if troubleshooting is strictly limited to *formal* logical inconsistencies can Framework Dependency be denied. Popper's philosophy of science might seem to comply with such a restriction, because the only kind of scientific problem it acknowledges is a theory's supposedly purely formal predictive failures.[6] But it is doubtful whether such a restriction can be viably maintained even with regard to science; whether the full range of scientific predictive failure can be accounted for formally in a manner that transcends *material* conceptual content. Outside science, however, in the welcome and potentially self-correcting critical challenge posed by the normative diversity encountered within the Open Society, such an account is impossible. There, with regard to politics and ethics, the type of self-criticism Popper identifies with adopting a rational stance could hardly be restricted to questions of mere logical consistency. However, pursuing these questions any further would take us too far afield.

Enough has been said in the previous two chapters to indicate that the marked disinclination of philosophers of science committed to Framework Dependency to apply the principles of Critical Rationalism to frameworks, and that of critical rationalists to concede the constitutive role of frameworks, is a symptom of heartfelt difficulty rather than an accident of history or the result of philosophical oversight. Committing oneself to both principles is not in itself problematic, as long as one concedes Kant's own contention that the constitutive linguistic framework in question is fixed and unchangeable, or, as in the work of Wittgenstein, Sellars, Taylor, Frankfurt, McDowell, Brandom, Korsgaard, and many others, one limits the discussion of meaning, normativity, agency, and so forth to the confines of a given framework. None of these thinkers *denies* that the constitutive linguistic frameworks their accounts premise are modified and replaced over time. Yet at the same time, the work of none of them explores in any detail the dynamics of such transitions, or the possibility of conducting them rationally. In philosophy of science, however, the question cannot be so easily ignored. If one concedes Framework Dependency in the sciences, then *pace* Kant, no properly philosophical account of the dynamics of scientific development can ignore the fact that the sciences have all undergone radical framework transitions in the course of their history. Acknowledging their existence, even charting them in detail will not suffice. Such landmark events in the life of a science, when deemed to have been rational, demand *explanation*.

The Standard Picture

With a focus on their endpoints, such transitions are commonly envisaged as two-stage processes in which a small number of practitioners—very often just one—first fashion and make public a framework radically different from the one governing their field, and then, in the second stage, succeed in winning over a critical mass of fellow practitioners, who in endorsing the new framework transform the field. To explain the first phase is to understand the developers' reasons for seeking an alternative framework in the first place; to explain the second is to understand the community's reasons for preferring the new to the old. My main objective in this chapter and the next will be to suggest and argue for a more complex account of framework transitions than the standard two-phase picture. But it is important to stay with the standard picture for the time being, to appreciate the problems it raises for any viable alternative.

The first phase presents the most serious stumbling block. In some areas of human endeavor—especially, though by no means exclusively, in the arts[7]—fields are transformed by changes of style and genre that are initially introduced experimentally, even playfully, with an eye to challenging convention by thinking "outside the box," and later endorsed for their sheer novelty. But in areas such as ethics, political theory, theology, and science, we would like to think that radical change can in principle be introduced for the sake of more than its originality per se—that one's ethical outlook can be modified for ethical reasons, and a community's scientific framework for reasons scientific. For practitioners of a scientific framework to have reason—not an ulterior reason, that is—for seeking to replace it means they have found it sufficiently flawed or lacking *scientifically* to merit replacement.[8]

In other words, the problem of deeming the first phase of a framework transition in science rational (that is, of describing it as undertaken *for* a reason, not merely as justified in retrospect) comes down to the problem briefly described at the end of chapter 1, namely, that of articulating the possibility of practitioners adopting a normatively critical stance toward the very normative framework constitutive of their thinking—a problem justly considered insoluble to the point of incoherence. For how can a constitutive framework be impeached by the very principles of rational reasoning of which it is constitutive? How could a committed Newtonian, for example, argue for the need to replace his Newtonian framework by force, exclusively, of Newtonian argument? Normative self-criticism necessarily requires prior commitment to a framework of norms and standards, which, due to that commitment, is exempted in principle from the type of criticism it makes possible.

The second phase of the process, in which the community at large elects to endorse the new framework in place of the old, would seem less blatantly problematic than the first. Perhaps opting rationally for a fully developed alternative framework in place of one's own does not require deeming the latter scientifically wanting by its own standards, but only relatively so in comparison to the new. Michael Friedman's account of the rationality of scientific framework replacement is presented along such lines. I shall return to the second phase in the course of discussing Friedman's position further down. But it is the first phase that really counts. The need for a new and radically different framework must be found necessary, and an appropriate candidate must be thought up, largely developed, and made public by its framers, before the second phase can begin. The creative process of devising and arguing for the new framework is what sets such transitions in motion and lends them both their force and content. The second, communal phase notwithstanding, the process rationally requires that those responsible for fashioning and proposing the new have deemed the old scientifically wanting while still in force. And that, we have seen, seems quite impossible.

Without saying so explicitly, philosophers of science have generally treated the two principles—Framework Dependency and Critical Rationalism—as incompatible, as if they are forced to choose between them. Frameworks are immune to normative criticism from within, it is implied, not because they are fixed and unchangeable, but because they are *constitutive* of such criticism: They are criticizable, argue Popperians, and therefore lack constitutive force; they are constitutive, argue neo-Kantians, and therefore cannot be criticized. Although I know of no philosopher of science who has articulated the problem in so many words, one would be hard pressed to name a single one whose work displays serious commitment to the grounding ideas of both Popper *and* Kuhn.[9]

Even Friedman, with whose notion of the relativized a priori and deep concern for the rationality of framework replacement in science the present work is in general agreement, carefully avoids all talk of what might have rationally motivated the individuals responsible for the first phase of the process. He analyzes what was newly introduced and what of the old was retained in the course of the framework transitions he studies, but the intense erotetic processes of deliberation by which those involved deemed the elements they retained to be suitable, those they discarded wrong, and those they modified wanting is passed over in silence. Friedman does not raise the problem of squaring the two principles explicitly, but he does write *as if* it is insoluble; as if his firm commitment to Framework Dependency renders the keen pondering and earnest deliberation of the agents responsible for the moves he charts irrelevant, or external to the rationality of the transitions for which they were responsible.[10]

AIM AND STRATEGY

What follows is an attempt to argue against the grain for retaining serious commitment to both neo-Kantian principles described at the outset. Building on my former work, the remainder of the present chapter proposes a general account of rationality as normative self-criticism that remains fully committed (a) to the framework dependency of science in the strong sense of the term, late of Reichenbach, Kuhn, and Friedman; (b) to the rationality in the above sense of framework replacement in science; and (c) to the radical diversity of successive scientific frameworks (which Friedman sees fit to play down).[11] By critically developing two of Friedman's most novel departures from Kuhn, chapter 4 explores and develops the historiographical implications of such an account, which will provide the narratological guidelines for the historical case study presented in the chapters that make up part 3. But first, let us turn to the general question of normative self-criticism.

My aim is not to contest Friedman's account, so much as to *complement* it with an intrasubjective account of scientific agency, which, as I shall argue, necessarily presupposes an intersubjective context, both of which are in a sense implied, yet remain unacknowledged, in Friedman's work.

External Sources

The first of Friedman's departures from Kuhn is to insist that the physics community is not, and cannot be, the only source of ideas capable of proving relevant to physics. Key elements eventually incorporated in new scientific frameworks, he shows, were first discussed at length in discursive settings on which the prevailing scientific framework had no constitutive hold, long before they were put into play by physicists. This is an important observation, and not only for its historical accuracy.[12] Serious consideration of alternatives to a truly constitutive framework requires an incentive and ability to think out of the box that one does not expect to find among practitioners for whom that framework is constitutive of what "counts as really (physically) possible."[13] The very idea of a truly constitutive, yet, at the same time, rationally replaceable, framework requires, according to Friedman, external sources and arenas of unconventional thinking.[14] In almost all his published work to date, including his detailed contribution to Domski and Dickson 2010, Friedman locates the external source of such thinking exclusively in the fields of mathematics and scientific philosophy. Only in Friedman 2010b is it broadened to include other areas of thought—such as theology. I shall return to this important point further down. But first, to pinpoint our disagreement.

Friedman implies that an unconventional idea's mere *availability* to scientists attentive to these fields renders it a viable option for them, which in turn explains the rationality of their taking it seriously. While the "constitutive principles operating at a given time and in a given historical situation define . . . for all practitioners of a given paradigm . . . what now counts as really (physically) *possible*," the "succession of new mathematical, empirical and philosophical situations . . . makes it *intelligible and reasonable* for those operating within an earlier paradigm . . . to entertain—seriously to entertain—a transition to a later one . . . to envisage—seriously to envisage—a *genuine expansion of the space of available intellectual possibilities*."[15] Or, as he elaborates the same point elsewhere:

> Although Einstein's theory remains conceptually incommensurable or nonintertranslatable with the Newtonian theory it replaced, the way in which it was inextricably entangled with the previous traditions in question explains how it was nevertheless rational . . . to accept Einstein's radical expansion of the Newtonian space of conceptual possibilities as both mathematically and physically possible (as, in Kantian terminology, *really* possible).[16]

Einstein's radically new conceptual framework, he adds further down, not only *expanded* the Newtonian space of conceptual possibilities, but "in fact evolved from" the Newtonian framework it replaced, "through precisely an intervening tradition of mathematics and scientific philosophy."[17]

But the fact that philosophers or mathematicians have happened to toy with an idea that flies in the face of scientific sensibilities—non-Euclidean geometries, for instance—is, of itself, no reason for any scientist to seriously entertain it *scientifically*, let alone go out of his way to incorporate it in a near-meaningless modification of the framework constitutive of his work and thinking. For Poincaré to be deemed rational in considering the *scientific* viability of a non-Euclidean form of intuition adopted by convention requires, I insist, that he do so *for a reason*, namely, in response to a felt failing or inadequacy on behalf of the framework still in play. If we consider it impossible to take such a self-critical stance toward the framework constitutive of one's own work, then it matters little where and how the renegade idea originated, or why the agents who elected to run with it might have initially found it attractive.

DiSalle's Challenge

As noted briefly above,[18] Robert DiSalle, who otherwise fully concedes Friedman's position, has rightly called him to task on precisely this point, urging him to "reconsider the role of philosophy in the evolution of science—not merely as an external source of general heuristic principles and new conceptual possibilities, but, at least in the most important revolutionary developments, as an objective tool of scientific inquiry."[19] In a series of publications,[20] DiSalle argues that, in addition to the role attributed to philosophy by Friedman as a source of new conceptual *possibilities*, the history of the transitions from Aristotle to Einstein exhibits significant moments in which the physicists themselves engaged in keen "dialectical confrontation with prevailing conceptions,"[21] with a view explicitly to "revealing the hidden presuppositions of the old conception, and exhibiting the internal difficulties that must be resolved by the new."[22] He portrays physicists as not merely alert and open to "the rational discussion of theoretical alternatives"[23] among philosophers and mathematicians, but actively partaking in that discussion as part of their work as *physicists*. "The role of philosophy in conceptual transformation," he urges,

> goes beyond its contribution to enlarging the common conceptual space, and providing for intelligible discourse across theoretical divides. In the most important cases . . . we find that philosophical argument is essential, not only to the rational evaluation of a new theory, but to the very construction of the theory; the conceptual transformation is precisely the product of a conceptual analysis.[24]

Like Friedman, DiSalle makes his case historically by showing in considerable detail how Galileo, Newton, and Einstein all fashioned their conceptual innovations in response to shortcomings they perceived in the frameworks prevailing at the time. In doing so, he raises two related metahistorical issues highly relevant to the concerns of the present study. The first concerns the question of what it is for a framework transition to be deemed rational. His analysis clearly implies that *two* questions of rationality need to be addressed, pertaining to the two-stage picture referred to above: that of the creative individuals who in the first stage propose to modify or replace the framework governing their work, and that of the rest of the community who, when faced with their new suggestions, elect in the second stage to pledge allegiance to the new. DiSalle's work clearly implies (though nowhere states explicitly) that while the second form of rationality is, as Friedman maintains, essentially one of retrospective, comparative (and communicative) rational *choice*, the first, wholly lacking from Friedman's account, must turn *prospectively* on an agent's ability to (noncomparatively) fault elements of her own constitutive framework.

The second, more philosophical issue implicitly raised—though, again, nowhere addressed by DiSalle—is, of course, that of the very possibility of adopting such a critical stance. As I have argued, this poses an exceptionally thorny problem, especially from the neo-Kantian perspective he supposedly shares with Friedman.

DiSalle's suggestions are an important complement to Friedman's picture, which the latter justly, if not quite wholeheartedly, welcomes.[25] Perhaps Friedman's hesitancy signifies an uneasiness on his part with the total nonchalance with which DiSalle introduces his protagonists' ability to criticize the frameworks constitutive of their thinking. The present study may be seen as an attempt to take seriously the centrality DiSalle attributes to conceptual analysis in framework transitions by facing up to the philosophical problem, which, although it begs, DiSalle ignores.

The Rortian Alternative

The problem of stage one is not only thorny, it is pressing. If scientists cannot be described as responding to problems, as striving knowingly to mend, restore, or improve the framework they propose to modify or replace, then we have no choice but to view them, as Richard Rorty does, as aimlessly or ironically toying with the strange and the exotic with a view to stumbling inadvertently on an original alternative that may prove worthwhile in retrospect.[26]

I doubt this is the type of rationality we would be comfortable attributing to a Galileo or a Poincaré! Rorty, of course, does not find such a position in the least embarrassing:

> [The] Wittgensteinian analogy between vocabularies and tools has one obvious drawback. The craftsman typically knows what job he needs to do before picking or inventing tools with which to do it. By contrast, someone like Galileo, Yeats or Hegel . . . is typically unable to make clear exactly what it is that he wants to do before developing the language in which he succeeds in doing it. His new vocabulary makes possible for the first time a formulation of its own purpose.[27]

For Rorty, even the most significant scientific upheavals add up to no more than an uncritical process in the course of which practitioners, as he puts it, "gradually lose the habit of using certain words and gradually acquire the habit of using others."[28]

As DiSalle argues repeatedly, it seems absurd not to be able ever to describe those responsible for framework replacement in science as exercising their agency in self-conscious and creative response to a perceived crisis.[29] And yet, taking seriously the constitutive force of linguistic frameworks seems necessarily to deem such talk incoherent. *Because* they are definitive of their practitioners' standards of propriety, linguistic frameworks, it is argued, cannot be deemed inappropriate or unfitting *from within* by those whose frameworks they are. Friedman's account of the rationality of framework transition is motivated primarily by the problem of incommensurability—namely, by the problem, typical of the second stage, of arguing comparatively for a radically different, well-formed alternative to the framework from which one is arguing. Despite my reservations regarding his proposed solution,[30] the problem it addresses pertains exclusively to the second stage of the process, in which a fully-fledged alternative framework has already been made available. It is a problem categorically different from that of accounting for the rationality of the first stage, for the *reasons* an agent might have for wanting to develop such alternative in the first place.[31] Friedman leaves this latter question of the rationality of framework *construction* elegantly unattended.

RESTATING THE PROBLEM

But can the problem of the rationality of framework construction be met? I believe it can, but to see how, the problem needs to be further narrowed and more pointedly stated. As I put it at the outset, as much as one would like rational agents to be required (both in and outside science) not only to exercise their norms *in* self-criticism, but to make them the *object* of such criticism, no serious philosophical account of agency and normativity that I know of makes room for such feats of self-alienation—and we now understand why.

Put crudely, to concede Framework Dependency and the rationality of framework replacement, yet reject the idea that the latter requires normatively faulting the framework from *within*, is to conflate scientific reasons with some form of ulterior motive, or at least reasons external to science. Carnap, the later Kuhn, and Friedman all concede Framework Dependency in the strong sense of the term, yet ground inter-framework rationality in factors they believe are *independent* of the framework in question. Carnap stated it the most clearly in claiming that "external questions"—questions concerning which scientific linguistic framework to adopt (as opposed to "internal questions" that are decidable by means of the "logical rules" of an already adopted framework)—could be decided only conventionally on the basis of pragmatic, extrascientific considerations.[32] Kuhn's work dithered between his well-known likening in *Structure* of paradigm shifts to a-rational conversions akin to gestalt switches,[33] and a more universalist positing of puzzle-solving criteria, such as accuracy, precision, scope, and simplicity, that supposedly remain permanent across inter-paradigmatic divides and are therefore capable of rationally facilitating revolutionary scientific change.[34] Friedman, who firmly rejects both these options,[35] opts for a third, according to which ideas radically at odds with the prevailing framework are rendered "live options" and hence become rational possibilities for its practitioners for no better reason than having been seriously considered by members of other disciplines.

As I have argued elsewhere,[36] although I find Friedman's criticisms of Carnap's conventionalism, of the early Kuhn's gestalt-switch account of paradigm shifts, and of the later Kuhn's assertion of inter-framework instrumentalist criteria all to be decisive, I am unable to accept the alternative he offers. The mere articulation by others of elements of a framework radically at odds with one's own is not reason to endorse them. If framework *replacement* in the hands of its practitioners is to be considered rational, they must be portrayed as having had reason to do so, which necessarily requires them, *pace* Friedman to deem the framework constitutive of their scientific thinking to be sufficiently *scientifically* flawed to do so. But as we have clearly seen in considering the works of Korsgaard, Taylor, Frankfurt, and McDowell, commitment to the identification of rationality with this kind of critical reasoning necessarily commits one not only to the existence of such a constitutive framework, but also to its exemption from the normative critical scrutiny of those whose framework it is. The argument is quite general, and applies to any framework constitutive of normative criticism, scientific or other.[37]

The only conceivable way around this vexing impasse, we have seen, is by strictly restricting one's notion of scientific self-criticism to trouble-shooting for formal logical consistency—as Popper's philosophy of science appears to do.[38] But, as we have seen in discussing Brandom's Kantianism, this is an intolerably high price to pay, for the way it rules out any notion of scientific *propriety*. From a Newtonian perspective, phenomena such as wave-particle complementarity, quantum superposition, or the dependency of time, mass, and length on velocity do not merely *contradict* theory, they are scientifically *unfitting* to the point of incoherence. For a bona fide Newtonian practitioner to suggest they exist would be the equivalent of a player taking a jump shot for a goal in a soccer game! They are not merely inconsistent with Newtonian theory, but are in defiance of the very Newtonian norms of *scientificality*. They are therefore *scientifically* unthinkable, just as the idea of inertia was unthinkable in Aristotelian physics.

This does not stop imaginative philosophers or novelists from speculating about such eventualities or physicists from listening in, but being imaginable outside science does not make them any less inappropriate *for* or *in* science. To limit criticism to questions of formal consistency is to unacceptably drain the discourse of any notion of scientific *normativity* and normative commitment. Yet to retain a robust notion of normative self-appraisal commits one, as we have seen, to a dependency on a framework that is immune to normative criticism, and hence to rational assessment. Hence the impasse.

Cast in the terms of Frankfurt's powerful vocabulary of selfhood, what lends the I-self its special judicial status is that it is constituted not merely by what one earnestly wants, but by what one earnestly wants to want, and what one wants to want pertains to what one cares about most dearly. This is what makes one's I-self the seat of one's norms and standards, and of the "volitional necessities" that explain the normative hold they exert on one.[39] One's I-self is capable of self-*reflection*, of course, but not of self-*criticism*. It can judge an insistent desire or pressing impulse to be worthy or unworthy of identification and endorsement, but it cannot, of itself, normatively impeach any part of itself—which is just another way of saying that a person cannot change at will what he wholeheartedly wants. Standing back from our I-self, we are left with no normative evaluative means to hold it in check (other than to scrutinize for clarity or consistency). One can always circularly approve, but never normatively *disapprove*, one's own normative outlook. It follows that the problem of rational framework modification, scientific or other, cannot be addressed at the level of personal, intrasubjective normative self-criticism *simpliciter*. Left to her own devices, a person is incapable of her own accord of finding her norms normatively wanting.

Communal and Extra-Communal Contexts

As with intrasubjective normative self-criticism, the same holds true for *intersubjective* dialogue within groups, such as scientific communities whose members share a constitutive framework. However, as Brandom has argued extensively, in this kind of discursive setting an additional critical space opens that is unavailable to us at the purely intrasubjective level of deliberation. It is the space between what he terms "normative status"—namely, what a norm *is* (such as the meaning of a concept in a given language)—and "normative attitude," namely, what the norm is *taken to be* by a particular speaker.

Brandom's move is premised on a distinction between normative force and normative content. On the Kantian model, normative force—the force of one's commitment to a norm—is, in Brandom's terminology, attitude-dependent. According to Kant's autonomy criterion (which is adopted in various degrees of explicitness by all the thinkers we have so far considered), it is up to us whether we are bound by a norm. Normative commitment is *enforced* by an act of willful self-binding. It is *we* who make ourselves responsible. However, writes Brandom,

> if the *contents* of our commitments, *what* we make ourselves responsible *for*—were also up to us, then to paraphrase Wittgenstein, "whatever seems right to us would be right." In that case talk of what is right or wrong could get no intelligible grip: no norm would have been brought to bear, no genuine commitment undertaken, no normative status instituted.[40]

Judging the person signaling me to stop to be a policeman, or the brief contact between the players to constitute a foul, may be compatible with, and integratable into, the body of my other commitments. Subsequent information, gained from getting closer or watching a replay, may oblige me to withdraw and modify or replace my initial judgment. All of this is up to me. But the specific contents and consequences of deeming someone a policeman or something a foul are not up to me. In binding myself by making or retracting such judgments, I *become bound* to be committed by whatever they or their absence entail in a manner that is largely not up to me.

But where does the authority of these contents reside? If they are not up to us, to what or to whose authority do they owe? From the outset, Brandom's answer has followed the later Wittgenstein in firmly locating the source of normative authority in community practice. But in his *Tales of the Mighty Dead* (2002, referred to hereafter as *TMD*), the question of normative authority is raised for the first time from an angle closer to the concerns of the present study. His answer, briefly sketched (though richly exemplified) there and rendered more fully explicit with the publication of his Woodbridge Lectures in 2009, signals the additional Hegelian turn his thinking has taken in recent years. Still, Brandom nowhere addresses the problem of rational framework replacement as such, and as it stands, his work offers little toward properly resolving it—mainly because, as we have noted, the notion of self-criticism he associates with rationality is so limited. But it does suggest, or at least provides the groundwork for, a way out. Let me, therefore, briefly summarize this development in the role of the social in Brandom's account of normativity.

The opening chapter of *TMD* briefly discusses five nested "Conceptions of Rationality,"[41] the last two of which—the "inferentialist view" at number four and the "historicist view" at number five—jointly represent Brandom's evolving position at the time. The former represents the inferentialism of his *Making It Explicit* (1994, hereafter referred to as *MIE*), and *Articulating Reasons* (2000), in which rationality is identified with engaging others in giving and asking for reasons with reference to an existing set of socially grounded norms. The "historicist conception of rationality" represents the new development in Brandom's thought that would seem to bring it firmly in line with the concerns of the present study, in purporting to venture beyond rationally applying a given set of norms and account for the rational formation of *new* norms. In it the synchronic dialogue of reasons, so central to Brandom's earlier paradigm of rationality, is replaced (or supplemented) by a different and wholly diachronic picture of intersubjective encounter modeled after Hegel on the judicial development of common law. In *TMD*, the two images are retained side by side as representing two distinct levels of rational discourse—reasoning *with* a given system of norms, and reasoning *for* expanding it. The former, as I argued in chapter 1, suggests an important, if unacknowledged, element of normative criticisim, which, as we shall see, is wholly lacking in the latter.

In Brandom 2009, the most developed version of his position to date, an additional Hegelian twist is lent to the first level,[42] but the image of deontic scorekeeping and the game of giving and asking for reasons, and with them any suggestion of normative criticism, are dropped completely. From the point of view of the present study, this is a most unfortunate development, but if I am right in assuming (as indicated in chapter 1) that Brandom never properly appreciated the rational significance of the game of reasons in the first place, it was perhaps to be expected. Following Brandom through in this regard will bring us to the heart of the matter, and subsequently to what I believe is a possible way out.

Central to the argument of *MIE* is the idea that although rational agents make themselves responsible for their normative choices, they are nonetheless liable to misrepresent or misapply the norms they employ. Hence the distinction between normative attitude and status. Equally central to *MIE*, as I have noted more than once, is the idea that engaging in the game of giving and asking for reasons is definitive of rationality. Brandom makes much of both these points in that work, but fails to relate them in any way, unwittingly holding his definition of rationality hostage to his pragmatism. Rationality is defined by him in terms of what rational beings *do* and are capable of doing.[43] According to Brandom, those who engage in the deontic practices of giving and asking for reasons are rational *by virtue of* so doing. But even he admits that rationality consists first and foremost in *having reasons* for what one does. Only seasoned producers and consumers of reasons are capable of engaging in the game of offering and assessing them. If being a player in the game is to epitomize rationality, rationality's defining feature must be located *upstream* from the actual activity of playing, namely, in the reasons rational agents might have for playing the game. Rationality may well *assert itself* in, and may even be *warranted* by, participation in the public exchange and assessment of reasons, but if the players are rational, they will be engaging in the game *for a reason*, in which case rationality must be ultimately *identified* with *that* act of reasoning, rather than with its practical outcome. With respect to rationality, the pragmatist's arrow of explanation must be reversed: People are not rational because they play the game, as Brandom would have it; they play it because they are rational! Playing the game is a sure *sign* of sapience, and a dependable *indication* of rationality, but in *MIE* Brandom consistently mistakes the symptom for its cause. Here, as briefly suggested in chapter 1, I propose going a step beyond Brandom in this regard.

Undertaking to act rationally does not of itself compel or oblige one to expose one's reasons for so acting or to be curious about anybody else's. It is easy to imagine people who act for definite reasons they are reluctant to discuss with others, or people who are wholly indifferent to the reasons others might have for acting. As suggested in chapter 1, the only possible incentive for wanting to expose one's reasons to others and oneself to theirs that merits deeming doing so the very epitome of rationality is *self-doubt*. To act for a reason, we have seen, requires taking critical stock both of what one is considering to act *upon*, and of one's incentives for so doing. For a person *S* to have reason for doing *A* to, or in, situation *B* requires claiming minimally that in *S*'s opinion, (a) *B* in some arguable sense requires *A*-ing, and (b) that it is also in some arguable sense up to *S* to do so. In doing *A*, *S* announces the conclusion of her initial deliberations on the matter.

But, I propose, in going on to subject her reasoning to the normative appraisal of others, she pronounces a wariness regarding the finality of her resolve, which, she implicitly admits, she is incapable of addressing alone. Becoming a player in the game of giving and asking for reasons is to engage others in helping one do what one cannot in principle do alone: effectively criticize the conclusions of one's own self-criticism. With the help of others, one hopes to establish a second-order level of reflective self-scrutiny with respect to one's twofold, first-order self-critical decision to act. Insofar as rationality is essentially self-critical, this is rationality at its utmost. For it is here, as I described it in chapter 1, that critical reflection comes full circle to engage itself. And the point at which it comes full circle is precisely the point at which intrasubjective, first-person deliberation exhausts itself, and is necessarily supplemented by intersubjective, second-person exchange.

Reasoning requires evaluative concepts by and with which to reason. Hence, reasoned action necessarily premises a given, up-and-running normative conceptual framework. And when one's norms are deemed to be given, the only kind of self-doubt capable of motivating such intersubjective intervention is doubt concerning their appropriate *application*. This is the connection Brandom fails to make. The critical space he opens between normative attitude and status (taken over from the meaning skepticism readings of Wittgenstein) is what can justify and lend meaning to the role definitive of rationality that he attributes (late of Sellars) to engagement in the game of giving and asking for reasons. Concerned about misapplying the concepts with which we reason, or of using them unaware of all they entail, we frequently engage others in order to gauge convention (as in the example of the policeman) or benefit from their greater authority (as in that of the foul).

However, apart from securing a significant and indispensable role for the essentially critical discursive settings of reflective endorsement, this kind of intra-framework calibration of normative attitude is, as it stands, largely irrelevant to understanding the dynamics of framework modification we are interested in here. The scientific equivalents of attempting to approximate normative status by acquainting oneself with and keeping score of other normative attitudes lie firmly within the confines of normal science and normal science training. Hence, by making explicit the implied self-critical dimension in Brandom's deontic scorekeeping, we have still not advanced beyond the bounded sort of "normal" reflection we encountered in the far more explicit (if wholly intrasubjective) accounts of agency proposed by Korsgaard, Taylor, Frankfurt, and McDowell.

But in *TMD* Brandom purports to venture beyond the rationality of properly applying the norms to which one is already committed to that of norm-*making*. He likens this form of rational endeavor to that of a judge in a common law tradition who

> justifies her decision in a particular case by rationalizing it in the light of a reading of that tradition, by so selecting and emphasizing particular prior decisions as precedential that a norm emerges as an implicit lesson. And it is that norm that is then appealed to in deciding the present case, and is implicitly taken to be binding in future ones. In order to find such a norm, the judge must make the tradition cohere, must exhibit the decisions that have actually been made as rational and correct, given that the norm she finds is what has implicitly governed the process all along. Thus each of the prior decisions selected as precedential emerges as making explicit some aspect of that implicit norm.[44]

At first blush, this form of intersubjective engagement seems far more complex and dialectical than the giving and asking for reasons that typifies the first level. The reciprocal dependency of the fruits of one's rational reconstructions and the normative statuses they are meant to establish would seem not to map easily onto the more strictly linear inferential orderings of commitment and entitlement one seeks to make explicit by means of the rational discourse at level one—all of which is further, and more systematically, elaborated in the reformulation his position receives in his Woodbridge Lectures, which I have briefly discussed elsewhere.[45] But I prefer to stay with the earlier account for its better focus on the way individuals supposedly exercise their agency in rationally forming new commitments—a question none of the thinkers we have discussed so far even raises.

But despite its seeming promise, a closer look reveals several problems.

The first is epitomized by the word *rationalizing* in the passage just quoted. In engaging her predecessors, Brandom's common-law judge is not forging a new norm at all, but seeking justification for one to which she is already committed. Unlike players in the game of giving and asking for reasons, Brandom's judge is not wary about her decision, she does not mistrust her judgment, and consequently does not consult the tradition with the hope of being corrected. Unlike the mutually critical game of giving and asking for reasons, this form of intersubjective engagement is wholly (and admittedly) tendentious![46]

Brandom's two "conceptions" of rationality both involve a valuable intersubjective element that is wholly lacking from the other accounts of agency we have discussed. But the repertoire of reflective normative activity he purports to cover is as limited as theirs, almost to the point of irrelevance to the concerns of the present study. For just like the others, Brandom seems unable to envisage rational reflection except in relation to given and unquestioned norms, limiting rational concern to their proper application, at the first level, and their appropriate approval, at the second. So even when talking about norm-making, it is not about agents deliberating between different normative options, or wrestling with norms they or others deem to be problematic. As it stands, Brandom's account of rational norm-determination shows no interest at all in processes of norm-*changing*. It is not about finding one's commitments wanting, or suspecting that they might be, or being eager to have them contested by others. It is not even about *creating* new norms, or about rationally subjecting oneself to an existing norm, but about bestowing normative approval on norms *already held*.

This is all the more apparent in the light of the larger project to which Brandom devotes *TMD*, to expose anticipated elements of the inferentialism developed and systematized in his own work, presupposed in the writings of Spinoza, Leibniz, Hegel, Frege, Heidegger, and Sellars. For even he would admit that no one *becomes* an inferentialist by reconstructing an inferentialist tradition. By doing so, a person can at most justify *being* one! As ingenious and compelling a common-law judge's account of past rulings, or a Brandomian tale of one's mighty dead predecessors might seem, they are in both cases knowingly construed and crafted to *justify*, never to *challenge*, their author's normative commitments. In Brandomian retrospect, one does not justify one's normative choices by arguing that they survive the implied *critique* of others, but by showing that they are to a significant extent partly and positively anticipated by them.

In the Woodbridge Lectures Brandom, as we have seen, describes level-one rationality in terms of the ampliative and critical tasks undertaken intrasubjectively in synthesizing one's unity of apperception. Level two is described in analogous terms, also as a process by which "the judge is performing what is recognizably a kind of synthesis by rational integration."

> On the ampliative side, the judge is extracting material inferential consequences from commitments [of the adjudicators of previous cases] . . . And on the critical side, the judge is rejecting prior commitments that would be materially incompatible with the current decision—by *not* treating those decisions, or the considerations they turn on, as valid or binding precedents.[47]

The critical "side" of the process of synthesizing one's own commitments is self-directed. The objects of critical scrutiny are one's own candidate commitments, and one's understanding of those to which one is already committed. Here, however, in the supposedly more markedly Hegelian account of the role of intersubjective engagement in norm-making, criticism is wholly directed *away* from oneself. The commitment in question is "the current decision," a decision already made by the judge in question, who criticizes those of former judges for being "materially incompatible" with his own. What is gained from engaging the tradition is approval, but nothing is learned.

It is in this sense that Brandom's second-level rationality sits uneasily at odds with the implied critical nature of the role assigned to it in contexts of norm application. Brandom's blind spot with regard to the former obviously owes to his lack of appreciation of the latter. The problem is that it solidifies into a far too rigid fault line that is basic to the more recent formulations of his position, creating an overly strict division of labor between the intrasubjective and intersubjective realms of authority, responsibility, and possible impact.

Brandom's distinction between normative content and force—between what a norm says and implies, and the commitment it commands—is both categorical and important. Less categorical is the way the authority of and responsibility to each are taken by Brandom to map rigidly their discursive relevance. Deeming normative force the agent's sole responsibility stands to reason. We are capable of *forcing* others to act in certain ways, but cannot *commit* anyone but ourselves to so acting. The idea that authority with respect to normative content be ultimately granted to the interlocutors also stands to reason, for if not, as Brandom argues late of Wittgenstein, the very notion of normative commitment would lose all its force. And he is also right in insisting that to be effective, that authority must be recognized by the agent in question.

But Brandom conflates authority and *say*, or discursive relevance. In our interactions with others, we do not limit ourselves to commenting solely on the content of their norms. When we criticize them, it is not only for misunderstanding, misjudging, or misrepresenting a norm. We frequently criticize others for adhering to a norm we believe they understand, represent, and apply perfectly well! And in doing so, we do not deny that the authority for so committing themselves is ultimately theirs. We criticize them for misapplying their authority, for choosing wrongly to commit themselves. Whether or not they then choose to pay attention to our misgivings is, of course, up to them. But—and this is the important point—insofar as our normative criticism is capable of bearing on their decision, it will not be in violation of the autonomy principle, and hence has no bearing on the way Brandom maps out authority and responsibility. Nonetheless, subtle as it is, his system makes no room for this kind of intersubjective input.

At level one, Brandomian discourse is (implicitly) critical, but because conducted by discussants who all share the norms under discussion, only their proper application can and will be questioned. At level two, the norms in question are clearly no longer shared by all the voices one engages. But because Brandom insists that the authority of others is only recognized with regard to normative *content*, only the voices of those who adhere to the norm in question will be heard and recognized. On a Brandomian showing, the voices of the opposition necessarily go unheard and therefore unheeded.

If Brandom had not cast his net so widely, and if, instead of aiming at giving a general account of conceptual content and intentionality by talking of sapience and its discursive social setting *in general*, he had done so with reference to a more realistic plurality of particular, normatively diverse, linguistic communities of the kind his common-law judge model tacitly assumes but fails to awaken to, he might have been more sensitive to the potentially challenging presence of discursive partners who question not the content, but the very normative *worth* of one's normative preferences. But he does not. Nonetheless, his firm neo-Hegelian insistence on the indispensability of the discursive, social settings of agency opens a way for rationally transcending the seemingly necessary givenness of agency's normative precommitments.

The first major step in this direction requires us to recognize, *pace* Brandom, that agency's most relevant discursive settings are those whose members contest rather than share one's most heartfelt commitments. Given the centrality of self-criticism to the exercise of agency, it is only natural to attempt to locate the community's role in this respect in its ability to enhance and deepen the individual's self-critical capacities, rather than weaken them by bolstering his self-confidence. The value of others, one would think, lies in their otherness rather than in the extent to which what they do or say resembles or anticipates oneself. Brandom's work implies this for first-level rationality, but fails to make it explicit, and forgets it completely at the second level.

Options Further Narrowed

Friedman, we have seen, presents an unsatisfactory intermediate position,[48] insisting, contrary to the tradition he builds on late of Carnap and Kuhn, that normative diversity *between* communities holds the key to accounting for the rationality of framework replacement *within* the community, yet, not unlike Brandom, failing to view it in critical terms. He fully understands that the rational motivation for the need to replace a framework cannot be articulated in that framework's own terms—that it is impossible, as I put it in chapter 1, for a committed Aristotelian to argue for the need to replace his Aristotelian framework by force, exclusively, of Aristotelian argument. The rational dialogue of framework replacement, Friedman insists, must therefore somehow transcend the volitional necessities of the individuals or community whose framework it is. Hence, *pace* DiSalle, the importance of Friedman's point about the potential relevance to physics of debates and deliberations conducted outside the physics community by mathematicians, philosophers, and others.

But the point must be generalized and the emphasis placed very differently than Friedman has done. Indeed, no serious, deliberative human undertaking is pursued in a void. Such enterprises boast a plethora of discursive contexts—political, cultural, religious, economic, technological, intellectual, didactic, and pedagogical—in which their frameworks are discussed and pondered, and at times questioned in ways that those whose discourse they govern are quite incapable of doing. Friedman does not raise the point explicitly, but I believe that this is the idea that essentially motivates Habermas's notion of communicative rationality, warmly endorsed by Friedman (2001), namely, that a person is incapable of rationally modifying or replacing her *lebenswelt* of her own accord, but is liable to do so when in critical dialogue with adherents to different and conflicting lifeworlds. I doubt that the incentive to do so is entirely social, as Habermas implies, but that is a different matter. The basic idea is that one needs to transcend one's own perspective in order to think out of the box. But the problem of framework transition cannot be reduced to merely that of *thinking* out of the box. We are capable of acquainting ourselves with normative commitments that ground forms of life very different from our own. The problem of the rationality of framework transitions is not that of being able to *think* about such commitments and understand them (namely, that of "meaning incommensurability," as it is sometimes called),[49] but that of deeming them rationally *endorsable*. As historians of science, we have little problem understanding systems of scientific thought very different from our own. But the difference between *being an authority* on Aristotelian physics or Newtonian fluxions and *becoming* an Aristotelian physicist or Newtonian "fluxionist," or even seriously considering such an option, is categorical. It is the difference between understanding a normative system and being committed to it. To judge a significantly different framework from one's own to be a viable "live option" for oneself, to use Friedman's phrase, one must be in a position to normatively fault one's own framework—a disposition, we have seen, one is incapable of assuming of one's own accord.[50] To this end, I insist, contrary to Friedman, *the company of critics*, rather than a supply of unconventional conceptual possibilities, is indispensable.

However, although a constitutive framework can only be *questioned* from without, it can only be effectively *changed* from within, by the practitioners whose work and thinking it governs. Only physicists can change physics! Hence, the philosophical problem of accounting for the rationality of normative framework replacement narrows further to that of accounting for an agent's capacity to endorse and internalize normative criticism leveled at him by others. But given that practitioners are incapable of their own accord of deeming their own normative frameworks normatively wanting, *to what extent can they be stimulated to do so by trusted external critics?*

We all know that external criticism can be very effective, but to what extent can its transformative effect be considered *rational*? How can a trusted critic convince us of what we are incapable of convincing ourselves of, or prompt us to do what we are incapable of doing alone? What is it about normative criticism leveled at us from without that can set in motion an otherwise inconceivable, considered, thoughtful, truly transformative process of normative *self*-reflection? If framework replacement in science can be rational, these are the questions we must address.

Wittgenstein, who as a rule steers wide of all questions concerning the dynamics of language-game *variation*, raises the question of the reasonableness and effectiveness of criticizing people across language-game boundaries only once, to the best of my knowledge: in *On Certainty*, his posthumously published response to G. E. Moore's well-known defense of common-sense realism. He speaks for many in admitting only two possible outcomes when disputing across such fundamental divides. He terms such criticism "combat" (*bekämpfen*), and sees no third option between total failure, when "each man declares the other a fool and heretic," and unreasoned "persuasion," the term he uses for "what happens when missionaries convert natives."[51]

The reasoning behind Wittgenstein's position is simple enough. Rationally accepting or dismissing criticism is in most cases a straightforward matter of evaluating premises, inferences, and data. But *normative* criticism is different, for it purports to question what agents take for granted in exercising their agency, to challenge their commitments to what they *cannot help* but care about. Unlike ordinary criticism, normative criticism inevitably runs up against its addressee's "volitional necessities." In other words, for the same reason that renders normative *self*-criticism impossible, such criticism, when leveled by others, can never straightforwardly *convince*. Even when the argument's premises are deemed sound and its reasoning impeccable, if its conclusions are considered normatively unacceptable, it will remain ineffective. Impeccably sound arguments to an objectionable conclusion are at best deemed paradoxes and set aside for later consideration, and in the worst case, dismissed defensively without further ado. This is the logic behind both Wittgenstein's decisive conclusion and Kuhn's account of the way anomalies that fly in the face of the prevailing paradigm are ignored in the course of normal science.

The question of rational framework replacement, if it can still be considered a question at all, hence reduces further to whether (and if so, how) external normative criticism can effectively stimulate transformative normative self-criticism *despite being inevitably dismissed*. The very idea of a consciously dismissed argument having such an effect will strike most as downright absurd. If an argument is powerless to convince, how can its impact be deemed rational? The fact that we have reached such an impasse, many would claim, proves by *reductio* that Rorty and Wittgenstein are right in asserting that our normative frameworks mark the outer boundaries of rational consideration, rendering unfounded the very idea of their rational review and replacement.

THE DYNAMICS OF REASONED CRITIQUE

It is true that arguments critical of a person's heartfelt norms can never *convince*, but their disputative ineffectiveness has little to do with the quality of their premises or of their arguments. Normative criticism is rejected not because its logic is deemed faulty, but because its conclusions are deemed preposterous (or unthinkable, or repulsive)! This is the strength of normative commitment, but it is also its weakness. To see why, a brief word about criticism in general is in order.[52]

The Critical Argument

Prudent criticism is a particular kind of speech act: a reasoned and addressed call aimed at motivating its addressees to acknowledge and to respond to the problems or failings it claims to expose. Popper and his school view criticism as purporting to expose "problem situations," which they envisage as objective, freestanding features of an agent's circumstances. What harbor problems, one might say, abstracting from Popper, are *objects*—theories, instruments, institutions, artifacts, plans, procedures, and strategies, the kind of objects I have characterized as "goal-directed systems."[53] A problem is a system's failure to adequately serve its purpose, a malfunction, such as a theory's failed prediction. Thus construed, problems are facts—facts of a system's substandard performance. And for Popper, criticism reduces to a fact-finding test-report, the mere exposure of which is expected to move all involved to action. Heeding criticism is for him paradigmatic of rationality exactly because it is a simple matter of instrumental reason. For what is it to act rationally, if not to ensure that the means we employ adequately serve the goals or interests they are employed to achieve, and to act to replace them or to put them right, when they do not?[54] But the account of criticism offered by Popper and his school overlooks some of criticism's most salient discursive, personal, and most important, normative dimensions.

One way to see this is by distinguishing between testing and criticizing—a distinction that some readers are liable to find a little forced at first. One tests an object with a view to disclosing its flaws and shortcomings. As such, testing is, of course, part and parcel of criticizing. But the two notions are not synonymous. Not all testing is performed with a view to criticize, and not all test results, grave as they may be, amount to criticism. When a car owner subjects his vehicle to an annual checkup, when scientists run routine tests in the lab, or check through their calculations, or when a pilot runs through her preflight checklist, "criticism" does not seem the right term to use, even when such tests reveal something to be seriously amiss. Test results amount to criticism, I suggest, not when used to prove *something* wrong, but only when used, in addition, to prove *someone* wrong, or to warn *someone* off. When a physician calls a person's eating habits into question in the light of his blood work, she will be criticizing, but if all she does is recommend a new course of medication or a change of lifestyle, she will have been testing and mending, but not criticizing. Systems are tested, but only *agents* are criticized.

Criticism is thus more than an attempt to stimulate someone to attend to a problem by merely *pointing it out*, or advising, or even pleading with him to do so—as when urged by one's accountant to cut one's overhead. Criticism aims at provoking a person to do so by proving *him* to have been, in some foreseeable way, *wrong*. One cannot be criticized for something one could not have been aware of, or which one was powerless to stop. Criticism confronts its addressees with failings for which it deems them responsible, calling them to action by accusing them of substandard performance. Herein lies criticism's essentially *normative* import. Unlike testing, criticizing involves an inherent element of blame or reproach. It carries an implicit demand for *self*-reckoning. When called upon to attend to a system in response to worrying test results, one is being *made* responsible for putting it right. But when called upon to do so in response to criticism, one is being held in some way responsible for it going *wrong*. It follows that all criticism, even the most technical, involves an inherent and defining element of *normative* criticism. In addition to urging us to repair or abandon the system it attempts to prove flawed, criticism also calls us to *mend our ways*!

Criticism's normative element, however, comes in two significantly different varieties, which roughly correspond to Brandom's two levels of rationality. Much of the time one is chided for being inattentive, careless, or neglectful of what one *knows is right*. What is criticized in these cases is the way one applies one's norms, not the norms themselves. "Given the standards you abide by, this shouldn't have been allowed to happen," our critics argue. "Given those standards, it would be wrong of you to even consider this an option." "Given what you stand for, how could you remain silent?" Here the normative component of criticism amounts to holding one to one's standards, and calling one to task for not properly living up to them. In science this type of criticism belongs by definition to normal science: practitioners of the reigning paradigm criticizing fellow practitioners for ignoring, misunderstanding, or misapplying scientific standards that none of them questions.

The Problem with Normative Criticism

But at times, it is one's norms themselves that are criticized—one's very adherence to them, rather than the substandard *use* to which one puts them. And criticizing norms, it turns out, is a matter categorically different from criticizing the quality of their application. As we shall see, even the way the critical argument is leveled in this type of criticism is significantly different. To see this, however, a second distinction is in order: that between doubting and criticizing.

Criticizing presupposes doubt, but should be firmly differentiated from merely doubting. To register doubt is to raise the possibility that something *might* be wrong, while to criticize is to attempt to *prove* it. Doubting requires little more than applying a question mark, raising an eyebrow; to criticize, one is required to frame an argument. But not just any argument—prudent critics aim at more than making their case to their own satisfaction. Their objective is to prove to those they criticize that something is sufficiently amiss to require their attention. To achieve its transformative objective, it is not enough that criticism be merely stated. To have its intended effect, the argument must register and must be endorsed as *self*-criticism by those criticized—if not, it will have achieved nothing.

For their efforts to find their mark, critics must argue, therefore, not from their own, but from their addressees' perspective, emulating as best they can *their* worldview and values, mounting their case on premises *they* hold true and in ways *they* deem valid. Here lies the essential discursive and especially empathetic dimension of all prudent criticism: To win the trust of, and pose a viable challenge to, those it criticizes, criticism must be leveled with intimate authenticity from *their* perspective. This, I believe, is what motivates the great emphasis placed by Michael Walzer on the necessary connectedness of effective social criticism.

The need to argue from one's interlocutor's point of view, rather from one's own, would appear all the more true when criticizing another person's norms. As Wittgenstein observes in the passage quoted above, when arguing across language-game barriers with a view to challenging someone else's very form of life, it makes no sense at all to argue from one's own perspective. Hence the dilemma: To be mounted effectively, normative criticism must be leveled from the addressee's perspective. But that, as we have seen repeatedly, is impossible! There is never available a set of premises one's addressee is liable to consider true that can be shown, convincingly *for her*, to entail a denunciation of her norms. For normative criticism to rationally motivate those it addresses to rethink and modify their normative framework, it must be capable of transcending the volitional necessities that constrain those whose framework it is. But how is that possible, if to do so effectively, it must be leveled from within those very constraints? The problem of normative criticism turns out to be graver than we had thought.

Prudent normative criticism directed against genuinely heartfelt commitments is not only unable to *convince*, the problem is that it is impossible even to *formulate*. There seems to be no way to prudently mount a potentially effective normative critical argument!

How Prudent Can Prudent Criticism Be?

To address the problem, it is important to realize that normative critics are to some extent aware of the problem, or at least sense it. Let us imagine for a moment someone who knowingly employs, say, a thoroughly Frankfurtian picture of personal identity and normative commitment, and is well aware of the conundrum. Let us also assume that our imaginary Frankfurtian critic cares for the person whose norms she finds normatively questionable, which she regards as firm and solid components of his I-self.

Care for her addressee may take one of two forms of response, according to the importance she attaches to the normative issues at hand. One is tolerance. If she resolves, despite her disapproval, to put up with her addressee's conduct, refrain from speaking her mind, and grant him his way, it will be out of respect for what she knows he cannot help caring for. One expects tolerance to be a Frankfurtian's first option, not because adherence to Frankfurt's picture of personal identity renders one more morally sensitive to a person's right to the normative system of his choice or, as Popperians would have it, because it renders one particularly open and welcoming to the critical challenge of normative diversity.[55] The Frankfurtian incentive toward tolerance owes not to the discourse of rights or to the ethic of the open society so much as to the way Frankfurtians ground normativity in the nonnegotiable volitional necessities of agency. But even Frankfurtian tolerance has its limits. When a person's conduct is deemed to be *intolerably* wrong, even a Frankfurtian's respect for her addressees' volitional integrity and inability to consider acting differently is liable to give way to disapproval, dissent, and eventually explicit reproof.

These are cases in which our Frankfurtian critic deems tolerance to no longer be an option—which can only mean that she considers her addressee's conduct not merely to be normatively inappropriate, but also to run up against some of *her* heartfelt commitments. She opts for criticism rather than for tolerance, she realizes, because what she takes as her own volitional necessities leave her no choice. But a prudent Frankfurtian also knows that finding her addressee's normative convictions normatively intolerable bespeaks a real and unbridgeable volitional incongruity. She knows, in other words, that the normative repugnance she articulates in terms of *her* normative framework *cannot be translated into his*. She knows that from *his* perspective the problem she detects cannot be articulated as a problem at all.

Notice that the challenge of prudent normative criticism has little to do with the problem of the incommensurability or untranslatability of diverse frameworks as such. It is the far more specific and limited problem of the *inarticulability* from within the framework criticized of the normative *problem* that its critics claim it harbors. The kind of prudent critics of whom I speak do not seek to *convert* other people to their normative frameworks (persuade them, in Wittgenstein's sense of the term), but to prompt them rationally to recognize and take action against the serious normative failings they believe plague *their* frameworks.[56]

Our Frankfurtian critic knows, in short, that framing her critical argument from her perspective is as pointless and ineffective *qua* criticism as framing it from his is impossible. She also knows that however she proceeds and whatever she does, for her criticism to somehow register and be taken rationally to heart, she must address her addressee, as far as possible, in his own terms, on the basis of premises with which he can identify. She therefore reasons that her only chance of driving her point home and prompting him to reconsider is to do so from an *imaginary* perspective, which, though not completely his, and certainly not hers, poses *as if* it is his! She will try to frame her argument from a perspective as similar as possible to his, yet sufficiently (and subtly) different from it to enable her to make her case. And she will present the argument as if arguing from *his* perspective.

Quentin Skinner has interestingly shown how, to communicate their ideas effectively, innovative thinkers often feel bound to conceal their radical intentions, and rather than present their views in stark defiance of the accepted norm, prefer to do so by giving subtle and surreptitious new twists to the conventional vocabulary.[57] Something similar, I believe, happens in prudent and informed normative criticism. Well aware of what she is doing, our imaginary (and admittedly contrived) Frankfurtian critic manages to catch her addressee off guard by quite deliberately framing her critique as if arguing from his acknowledged perspective.

In real life, people usually do this unthinkingly, as a matter of course, because, whether we like it or not, our normative assessments of others—the expectations we frame, the disappointment and frustration we register in the light of their reasoning and conduct—emphatic and open-minded as they aspire to be—are ultimately conditioned by our own normative outlooks rather than theirs. When we are disturbed by someone's conduct to the point of protesting, we are measuring it by a normative yardstick we strongly feel *should* be his. We accuse him of betraying norms that are not his own. Our moral outrage owes to the fact that he fails to live up to the person we strongly believe he should be (or had believed he was). In other words, even if we are not as perceptive and calculating as our imaginary Frankfurtian, when openly and prudently criticizing elements of another person's wholeheartedly held normative scheme to his face, we will inevitably be arguing against him from a picture of his I-self that differs significantly from the real thing, as it were, but we will do so motivated by the deep conviction that it should be true.

Unwittingly or not, keen, honest, direct, and reasonably articulated normative criticism challenges its addressees with an explicit *argument* that, of itself, has no realistic hope of convincing, but comes accompanied by a picture implied by its premises of (the relevant part of) its addressee's I-self, which though true to a large extent, differs significantly from his own. Unlike our imaginary Frankfurtian, real-life normative critics seldom criticize in this way purposefully. Rather, they argue as if their arguments were capable of being endorsed directly. But this, in itself, is of little consequence. Our imaginary Frankfurtian does not enjoy any advantage as a critic by being aware of what she is doing. The result of normative criticism is always very much the same: a critical argument incapable of convincing those it criticizes, whose premises imply a meaningfully altered, yet otherwise largely true account of their I-self. Arguing from the political left, critics will surreptitiously attribute to their right-wing addressees certain liberal norms in order to make their case, while critics arguing from the right will tend to smuggle in just enough conservative value to make their point.

Normative Criticism's Transformative Impact

If normative criticism is inevitably dismissed, how can the picture conveyed by its premises still be retained? The answer is twofold. First, if we consider a critic trustworthy and informed, we may well deem his efforts sincere, even while rejecting his conclusions as wrongheaded. Rejecting the criticism of trusted critics does not undermine their trustworthiness. We are able to dismiss their criticism while fully endorsing it as conveying an honest portrayal of how they see us!

Second, while the first point is true of all criticism, it is especially true of normative criticism by virtue of an important difference. When an ordinary critical argument is rationally rejected, it is argued *against*, an attempt is made to *refute* it. And to refute an argument is to refute its premises, and with them *any picture they might imply or entail*. The picture conveyed by a rejected nonnormative critical argument resembles that of a recording device its addressees consider faulty and distortive. Such a picture will have no feedback effect, no self-critical import, even when deemed honest. If it moves us to action at all, it will only be to correct the *impression* it gives of us, not to correct *ourselves*. Arguing with our critics in such cases boils down to criticizing their criticism, countering their arguments with counterarguments.

Normative criticism is interestingly different. Normative criticism is not set aside because it is deemed *refuted*, it is set aside because its conclusions are deemed unacceptable. Normative criticism is disapproved, if you wish, not disproved; it is dismissed not because its *premises* are found to be unfounded, but, as I said above, because its *conclusions* are deemed to be preposterous, unthinkable, or repulsive. Hence, although our initial response to normative criticism is usually far more unyielding than our response to ordinary criticism, it is typically decidedly less *reasoned* than the ordinary case. When rejected, ordinary criticism is dismantled, while normative criticism remains intact. This is why the overall dismissal of normative criticism does not logically entail (for its receiver) an analogous dismissal of everything it premises, as in the ordinary case.

We deal rationally with ordinary criticism well within the boundaries of our space of reasons, endorsing and dismissing its arguments always for a reason. Normative criticism, by contrast, is dismissed because it runs up against the given, unquestioned normative entry points of the reasoning process itself—those that supply us with reasons without being entailed for us by others. The consequences of dismissing normative criticism therefore all lie downstream of our unyielding, unpremised, challenged norms. Upstream from them there is nothing but the rock-hard necessities of will. And thus, because normative criticism is left unrefuted, the picture of our normative identity conveyed by its premises can linger on, register, and be retained uncoupled from, and without being automatically discarded along with, the argument's conclusions.

Still, not to be formally and consciously refuted is still a far cry from being taken to heart and having an impact. To understand how the normative profile painted of us by trusted critics may succeed in rationally moving us involves an additional factor, which Frankfurt notes in passing in an earlier work:

> It is a salient characteristic of human beings, one which affects our lives in deep and innumerable ways, that we care about what we are. This is

> closely connected both as cause and effect to our enormous preoccupation with what other people think of us. We are ceaselessly alert to the danger that there may be discrepancies between what we wish to be (or what we wish to seem to be) and how we actually appear to others and to ourselves.[58]

We do not need to deem our critics' normative portrayal of us true for it to attract our attention in this way. It is enough that it is not considered disproved. Even though we instinctively dismiss their arguments as unfounded, their picture of us is liable to continue to resonate, alerting us to the discrepancies between how we appear to ourselves and apparently appear to them. To stretch the playback device analogy a little further, it is as if we are struck by discrepancies between a recording and how we imagine ourselves to look and sound, but without yet being sure if the recording device is to blame (as the analogy to the ordinary case goes).

In this way, hard-hitting normative criticism deemed prudent has the power to create a sense of inner discordance and self-estrangement capable of transforming our critics' criticism into *self*-criticism. This, I submit, differently from Brandom, is where the social context of heartfelt *inter*personal normative discourse can play a crucial role in creating the inner distance and perspective required for meaningful *intra*personal normative self-questioning—which, without defying them, does not respect the Frankfurtian fault lines of one's self's normal inner division.

Trusted normative criticism makes one aware of a normative profile one is capable of identifying as a different version of oneself, in the same way we identify even a considerably distorted recording or picture as being of us. There is no *reasoned* way for one to compare the two directly and to find the new version normatively preferable to one's own. That would require one's still functioning I-self to pass negative normative judgment on itself, by itself. The only properly *reasoned* judgment available to one at this point is to deem the new profile normatively inferior to one's own. If trusted normative criticism is to have critical, transformative impact capable of striking a middle course between all-out reasoned rejection and unreasoned conversion, it will have to be by means of a less direct route.

In Frankfurt's account of agency, unmitigated alienation and wholehearted commitment are not the only options available to the self. The self may well experience definite *mixed* feelings toward an object of love, be of two minds about it: loving and not loving it at the same time. In Frankfurt's term, a person may be not merely unclear or unsure, but truly *ambivalent* toward a final end or the activity required to achieve it. Faced with a normative profile he recognizes as closely, but not fully, resembling his own, a person may well become ambivalent with respect to the points of incongruity between the two. In other words, as a result of his critics' efforts, a person is liable to find his attitude toward a well-defined final end, norm, or standard changed and rendered ambivalent.

But to develop ambivalence toward a norm or standard is not yet to solve the problem one's critics raise. It is the self's equivalent of acknowledging and articulating it. It is not a *reasoned* response to our critics' reasoned efforts—this will only come later—but it is a sign of critical impact. To become ambivalent toward a formerly wholeheartedly endorsed norm is a form of passing critical judgment, as if to deem it volitionally problematic. But it is not a reasoned form of demotive judgment, because it does not involve the application of a standard. The norm is not demoted as a result of being measured or assessed in relation to, in comparison with, or in the light of other norms. Ambivalence, in this case, is the sheer result, one could say, of the two pictures being allowed to merge briefly in the mind, as if the profile presented by one's critics is momentarily *grafted* onto one's own, in a manner secured by the broad, recognizable areas of identity. In the first instance the normative incongruities are not adjudicated, but sit side by side, creating ambivalence by their very juxtaposition. It is a moment in which one's I-self does not sit in judgment, but merely contemplates the version it is presented with, trying it on for size, as it were, relishing the familiarity of its very strangeness, and managing to contain the incongruities as an ambivalent whole, rather than as jarring parts. This kind of ambivalence is a form of demotive verdict that is handed down not by an act of judgment, but by what we might term a nonadjudicative juxtaposition of incongruous commitments.

To become ambivalent toward a norm or a standard, toward the terminal value of a final end or a form of activity, is to deem them normatively undecided. This in itself does not necessarily render them problematic. Being of two minds about an object is not to declare the object flawed. What *is* rendered problematic is the self itself! A person who persists in a state of ambivalence, writes Frankfurt,

> is volitionally fragmented. His will is unstable and incoherent, moving him in contrary directions simultaneously or in a disorderly sequence. . . . his will remains obstinately undefined and therefore lacks effective guiding authority. As long as he is unable to resolve the conflict by which he is torn, the person is at odds with himself.[59]

He describes ambivalence as a disorder that demands urgent attention: "a disease of the mind"[60] that "tends to alarm a person and to mobilize him for an attempt at 'self-preservation.'"[61] The problem with ambivalence, one might say, is that it represents a significant weakening of a person's capacity to conduct himself rationally. Being of two minds about a norm or a standard renders all relevant judgment impossibly incoherent. If the necessities of wholehearted normative commitment provide the basis for a person's volitional rationality and hence the grounds for practical reasoning, then, by the same token, ambivalence generates *volitional irrationality*, which in turn obliterates his very capacity for agency in the affected areas. The deep and urgent need to free oneself of ambivalence is no less than an insistent self-rallying to reclaim our capacity to be rational.

When trusted normative criticism strikes deep, it is capable of creating ambivalence with respect to the norms it questions. However, the problem that the one who is criticized is consequently made to face and respond to is different from the problem his critics sought to convey. The problem with which *they* present him concerns the normative failings, in their opinion, of certain of his norms, standards, or forms of conduct; the problem *he* is subsequently made to contend with concerns his inability to function rationally in those areas. In a sense, his is the more pertinent.

For a person to free himself of ambivalence, writes Frankfurt,

> it is not necessary that either of his conflicting impulses disappear. It is not even necessary that either of them increase or diminish in strength. Resolution requires only that the person become finally and unequivocally clear as to which side of the conflict *he* is on. The forces mobilized on the other side may then persist with the same intensity as before; but as soon as he has definitely established just where he himself stands, his will is no longer divided and his ambivalence is therefore gone. He has placed himself wholeheartedly behind one of his conflicting impulses, and not at all behind the other.[62]

Frankfurt sees normative ambivalence as deeply problematic and exerting considerable pressure on the person to solve it by taking sides in the dilemma. "If ambivalence is a disease of the mind, the health of the mind," he writes, "requires a unified will."[63] It does not matter on which side of the conflict he decides to fall, as long as his decision is firm and clear. To solve the problem is to unify the will, and to that end one side is as good as the other. But the decision on which side to fall is not, and cannot be, arbitrary. A person can no more decide what he wills by simply wanting it than he can do so by flipping a coin. To solve the problem is to become "finally and unequivocally clear as to which side of the conflict *he* is on." This is a decision rendered by a person's I-self. One would think, therefore, that in the case of normative criticism from without, the dice will always be loaded in favor of the normative commitments of which the person's I-self consisted before being criticized. But that would be to gravely underestimate the disruptive effect of normative ambivalence.

When a person becomes truly ambivalent toward certain elements of his I-self, he ceases, by definition, to wholeheartedly identify with them. This is the sense in which ambivalence is a form of volitional demotion. When "ambivalated" by criticism with respect to a norm or standard, one's commitment to them is impaired to the point that they can no longer function in their original, self-adjudicating role. They lose their normative authority, and thereby their I-self status. By the time a person turns to attend to the ambivalence caused by his normative critics, his I-self will have come to resemble the undisputed *intersection* between the two partly conflicting profiles. In this way, his ambivalence toward the norms and standards his critics hailed as questionable will be addressed by an I-self that, in principle, should not be biased in favor of its former configuration.

The outcome may take a variety of forms. Frankfurt seems to allow only two: retaining one's full original commitment to the norms in question, or endorsing one's critics' position and alienating oneself from them. But there is always a range of intermediary forms of retaining partial commitment toward certain aspects or components of the object of one's original love. The important point is that there is no reason for any of these outcomes not to be as rational and reasoned as human reflection is capable of being. Once the norms questioned no longer form a part of one's I-self, the conflicting tendencies and motivations they now represent will be observed as it were from a distance, objectified and evaluated by one's new and leaner I-self, similarly to any other first-order candidate. It is in this indirect way that insistent, trusted, and thoughtful normative criticism is capable of setting in motion a process by which a fully fledged, wholeheartedly endorsed component of its addressee's I-self can become temporarily dislodged, lose its privileged status, be stripped of normative authority, and become freely and impartially resubjected to the reflective scrutiny of his innermost volitional core.

Can more be said about how ambivalence can be resolved? In science one would like to think, perhaps, that when truly ambivalated, practitioners have recourse to what Kuhn came to see as the thin set of scientific values, such as simplicity, precision, and so forth, that supposedly remain constant across framework transitions. I think not. I fully agree with Friedman's rejection of Kuhn's suggestion that such a set of über scientific standards exist at all. More important, the resources available to an agent in dealing with her ambivalence are forever that part of her particular I-self that has remained intact, and nothing more general than that can be said. Even if there are norms to which all humans or all scientists happen to be committed, they will not be reached for by virtue of their universality, but only by virtue of the particular agent's commitment to them. That, I believe, is all that can and need be said.

None of this, however, can be accomplished alone. When we are left to our own devices, our I-self is incapable of splitting and subjecting a part of itself to the critical scrutiny of the other part. We are capable of becoming ambivalent, but not of ambivalating *ourselves* toward certain of our norms and standards. We cannot cease caring or loving at will. Ambivalence is a state we can find ourselves in, but cannot create voluntarily. When we are left to our own devices, the self-critical fault lines of the reflective self are beyond control and remain rigid and unchangeable from within. But trusted criticism can change them for us. When the rich intrapersonal dialogue of the self is set within the context of an equally rich interpersonal critical dialogue with others, effective transformative normative self-criticism becomes a real option.

To return one last time to our little thought experiment, our imaginary Frankfurtian, who knows all this, will conduct herself, therefore, not only as an effective normative critic of others, but also, and perhaps mainly, as a knowing *consumer* of trusted criticism. For she is fully aware of how limited her ability is for unassisted normative self-questioning. She will therefore *actively seek* the critique of people whose insights and motives she trusts, because she knows that only thus is there a chance that the grip by which her norms and commitments bind her might be loosened sufficiently to allow *her* to subject them anew to keen normative review from within. Given Frankfurt's attitude to ambivalence, the idea of actively seeking the disruptive, ambivalating challenge of normative criticism will undoubtedly sound very strange in the first instance. But if there is any truth in the Frankfurtian expansion of Frankfurt's own picture proposed here, there is an important sense in which temporary ambivalence can be deemed a valued and sought-for state, and therefore set as a desired end.[64]

External Criticism Real and Imagined

Does such criticism have to be real, or can it be imagined?[65] In both our private and professional lives, we constantly test ourselves and plan our moves by conducting thought experiments in which we toy imaginatively with different circumstances, scenarios, and discursive settings and partners. Think how we prepare ourselves for an important job talk or negotiation. (Much more of our practical reasoning is performed by imaginatively running and comparing scenarios than by mere deduction. And there is much to say for the way we play them out and how and what we learn from them.) Scientists, too, are frequently called upon to imaginatively anticipate and respond to the doubts and queries of people committed to different frameworks than their own, as when working on a grant proposal, preparing a talk to be given abroad, or writing a textbook. There is certainly much self-critical potential in these kinds of exercise, which often require them to articulate basics that "go without saying" in their home communities, and to explain and defend them in a language they believe their imagined audiences will understand.[66] So perhaps the line I draw between normative self-criticism and normative criticism leveled from without is less rigid than I make out, and perhaps DiSalle is right in claiming that self-questioning from within can indeed set a framework transition in motion?

I think not. There is a world of difference between imagining a dialogue and actually engaging in one. It is the difference between talking to ourselves and talking to others. The hold of our constitutive frameworks controls our ability to imagine their responses and criticisms as much as it limits our ability to think for ourselves. We all remember the mistakes we made when writing our first proposals, giving our first talks, and writing our first textbook. (Think of how differently Judaism and Christianity responded to their distorted images of each other than when engaging one another directly.) We get better at anticipating how others think and respond, not by writing more of the same, but by *gaining experience* of real encounters with financing committees, foreign audiences, and new students.[67] Imagined encounters begin to be truly effective only when they are built on the *memory* of real ones.[68]

The difference between an imagined challenge and a real one is real, if only because our ability imaginatively to stand in our opponents' shoes, see the world and ourselves in it as they do, and faithfully emulate their counterarguments is severely, if naturally limited by our own convictions and frame of mind. At the same time, however, our conditioned frame of mind and heartfelt convictions also often impede our ability to hear a real external critic. We often misunderstand their arguments and misinterpret their queries simply because we are unable to frame them, ourselves. But with real critics, if we are not overly defensive, the conversation can continue, and our misconstruals of their position can be detected and corrected in ways that are impossible when talking to ourselves.

As Kant has taught us, to be normatively compelled in the full sense of the term is to comply with a norm or standard, knowingly aware of the possibility of breaking it. This Kantian element of normative *self*-necessitation, enlarged on and expanded by Frankfurt and other latter-day writers, marks the point where autonomy and necessity meet and merge. It is a semi-oxymoronic moment of *self*-compulsion, of necessity that is *willed* against a real possibility of disobedience, and awareness of getting it wrong. Though not normally a *reasoned* moment, and certainly not a critical one, it *is* a moment of at least fleeting reflective self-distancing, in which our freedom and responsibility are briefly glimpsed and exercised in declaring "I must." And when, confronted by knowing and trusted criticism, such moments may become permeated by a deeper sense of self-estrangement, and the mere prospect of disobedience be juxtaposed by a real alternative, then, I believe, the possibility of unquestionable commitment giving way to reflective, normative self-questioning becomes a real and a rational option.

If such an option is deemed viable, then it is possible to take Framework Dependency with utmost seriousness, and still retain a notion of normative self-criticism sufficiently robust to sustain the idea of rational framework review and replacement as a third, deliberative option between Wittgenstein's two. The argument I have outlined is quite general, and should, in principle, apply equally well to science. If the transition to science can be made to stick, two important causes will have been served. First, it would no longer be necessary to make a special case for scientific rationality (or for that of mathematical physics), as Friedman does. Second, and more important, the wishful thinking of Popper and McDowell would be made a philosophically viable reality, even for those, like myself, who are firmly committed to the neo-Kantian grounding ideas of the dynamics of reason. This is the task of the chapters to come.

CHAPTER FOUR

Toward a Narratology of Scientific Framework Transitions

INTRODUCTION

The two preceding chapters detailed the two main ideas late of Kant on which this study rests—(i) a strong form of Framework Dependency, and (ii) an insistence on the centrality of self-criticism to rationality, and on its inherently normative dimensions. These, we saw, jointly lead to an insurmountable impasse when combined to account for the rationality of framework transitions. In philosophy of science the impasse manifests itself in a broad, if largely unspoken agreement that the two ideas are mutually exclusive,[1] with one school conceding Framework Dependency but denying that Critical Rationalism applies to scientific frameworks, and the other conceding Critical Rationalism to the denial of Framework Dependency altogether. Neither position, however, is as coherent as it might seem.

On the "Kuhnian" side, the rejection of (ii) in favor of (i) commits one to one of two equally untenable positions: to deny the rationality of framework transitions altogether, viewing them, as did the early Kuhn, as on a par with gestalt switches and religious conversions; or to deem them rational, but define their rationality in ways wholly devoid of normative self-criticism, as Friedman proposes. This second option, though an important step up from Kuhn, we have seen, is limited by Friedman to the second and less interesting phase of such transitions—when a new framework is already in play, and practitioners are faced with a choice between staying with the one to which they are committed or opting for the new—to the untenable exclusion of the first phase, in which a need for a new framework is felt and acted upon by the framers of the new. Nonetheless, as mentioned above and argued elsewhere,[2] even when limited to the second phase, Friedman's notion of rationality (late of Habermas) remains untenable for lack of any element of normative self-criticism, for, as we have seen, the very notion of acting for a reason inherently contains such an element. Phases one and two are no different in this respect. To have reason for seeking a viable replacement of one's framework, or for replacing it with a given, full-blooded alternative, is to deem it sufficiently flawed to merit replacement. Even if one limits the discussion, as Friedman ends up doing, to the problem of rationally deeming the new framework a "live option" for practitioners of old, it is senseless to do so without some criteria for judging the old normatively inferior to the new.

The Popperian option fares no better in this respect. Just as acting for a reason necessarily presupposes taking a normatively self-critical stand, taking such a stand, we have seen, necessarily presupposes a given normative framework, and hence inevitably commits one to Framework Dependency. The only way to steer wide of the kind of neo-Kantian form of Framework Dependency advocated by Kuhn and Friedman and conceded in the present study is to limit one's notion of normative scientific self-criticism to norms so general as to be deemed formally definitive of all scientific research as such. In identifying a theory's testability with empirical predictive failure, and deeming that the sole criterion of scientificality, Popper and many of his followers would seem to have achieved just that: limiting normative self-criticism in the sciences to troubleshooting instrumentally for formal consistency between theoretical generalization and singular observation report. But even within the Popperian camp, several thinkers criticized this form of fallibilism as being far too stark and too broad a criterion to account for such aspects of scientific rationality as the relative weight a community attributes to certain predictive failures in comparison to others, and the kind of diversity of commitment responsible for the development of distinct scientific research programs.[3] In taking such aspects of science into account, one recognizes notions of scientific normative commitment that far transcend logical consistency. Yet, as we have seen, identifying scientific rationality with making normative commitment subject to the test of instrumental success significantly undermines the seriousness of such commitment, rendering even their superior accounts equally untenable.

The previous chapter then went on to argue that the impasse can nonetheless be breached if a third idea is introduced, in addition to the first two: a "Hegelian," intersubjective corrective, which follows, yet in an important sense also departs from, Brandom's suggestive model. To recall, Brandom's account turns on a fundamental distinction between normative force and content. In Kant's system (as Brandom reads it) the autonomy principle is served by rendering normative force wholly and exclusively attitude-dependent, while normative content—that by which the actor is bound when binding himself by the norm—is rendered wholly attitude-independent by viewing the categories and forms of intuition as fixed and given independently of human comprehension. Brandom's Hegelian corrective to the idealism of the First Critique is to insist that even normative content be wholly attitude-dependent, though independent of the actor's own attitude. Normative content is established objectively for each individual speaker by the authorization he grants other speakers to hold him responsible to the norms for which he has made himself responsible.

> Hegel's reciprocal recognition model stems from the idea that accepting the overarching Enlightenment commitment to the attitude dependence of normative statuses, the way to make sense of the independent counterbalancing (meta-) authority associated with the content to which I commit myself or for which I make myself responsible, is to have authority administered by others, *to* whom I make myself responsible, *by* authorizing them to hold me responsible for the content I have exercised my authority to make myself responsible for.[4]

Brandom characteristically takes linguistic norms as his paradigm (Brandom's term).[5] One commits oneself by using the word *copper*, subjecting oneself to expert judgment regarding aspects of copper of which one may be ignorant, such as its specific heat, and those about which one may be mistaken, such as its melting point. But these are the easiest cases, for it is not one's norms so much as one's knowhow in applying them that is challenged. A little less straightforward are cases in which what is challenged is not the content of one's judgment, but one's judgment itself—when experts rule the metal not to be copper at all. Here the neat distinction between force and content is no longer so easily maintained, nor is the autonomy principle so singularly in force. For it is not the *content* or application of a conceptual norm with which one has bound oneself that is challenged and determined by others, but the correctness of one's very binding.

Still, when viewed from an inferentialist perspective such as Brandom's, criticizing a person's use of descriptive terms like *copper* remains safely within the unproblematic boundaries of criticizing his application of a shared vocabulary. But the discourse of commitment extends far beyond the realm of the descriptive—so much so as to render highly problematic Brandom's hailing of the "copper" example as "a paradigm of discursive (conceptually contentful) norms." The kind of criticism completely absent from Brandom's account is that leveled *across* such discursive divides, as when the normative outlook that grounds one's descriptive vocabulary is questioned. In the scientific context, Brandom's omission pertains to the difference stressed by Carnap, Kuhn, and Friedman between what the former dubbed internal and external questions. Brandom explicitly follows Quine in rejecting Carnap's two-tier picture of the logical structure of language as a "completely unrealistic" account of natural languages.[6] As a firm adherent to Framework Dependency, I naturally disagree, but because the present study focuses on science, I shall not press the point further with respect to language in general.

With regard to science, suffice it to say, as noted in chapter 2, that I follow Friedman in maintaining that it is possible to concede Quine's well-known rejection of Carnap's contention that the second-order rules of a language's constitutive framework can be formally distinguished from the first-order rules they determine, yet still to follow Kuhn in fully conceding Framework Dependency. It is not their formal necessity, but their constitutive function, wholly ignored by Quine and later by Brandom, that categorically distinguishes, say, the rules of the calculus or Newton's three laws from the rest of Newtonian physics.

Brandom's Hegelian notion of mutual recognition is powerless to account for the kind of rational normative realignment characteristic of scientific framework transitions, because as it stands, it gives no account of the transformative impact intersubjective normative criticism can have on heartfelt intrasubjective normative commitment. The preceding chapter proposed such an account by driving a wedge between the prior recognition of one's trusted critics' authority with respect to one's norms, which is seldom granted, and the ambivalating effect their criticism can have despite being rejected. Based on that account, the following pages argue for a narratology of rational framework transitions in science capable of accounting for both phases of the process without denying rationality's inherent self-critical constituent, and without having to make a special case for science—the two major shortcomings that plague Friedman's account.

Two Tasks

The site of rational normative realignment, the previous chapter argued, is the arena of intrasubjective deliberation; the ultimate discourse of rationality is that of the self. This in itself should come as no surprise. Insofar as "rational" is taken, as it is here, as denoting a category of reasoned or considered action, it stands to reason that personal commitment can only be modified or replaced *rationally* by the person whose commitment it is. But its two further conclusions are perhaps less unsurprising. The first is the absolute dependency that emerged of such intrasubjective rational deliberation on an intersubjective environment of trusted normative criticism. The second is the idea that the transformative high moment of rationality is not one of inspired discovery, but one of deeply disturbing ambivalence, a moment of inner discordance fraught with tortured dithering and painful indecision. Ambivalence, self-estrangement, and indecision are as a rule not the most inspiring and motivating states of mind, but in moments of normative crisis they become, I have argued, reason's defining disposition and very driving force.

From the neo-Kantian perspective I share with Friedman, such criticism is clearly not forthcoming from within the normal, paradigm-governed discursive settings of one's own field (a point to which, as we have seen, DiSalle's otherwise insightful critique of Friedman remains oblivious). Hence, with all due respect to the importance Michael Walzer assigns to the "connectedness" of effective normative critics, if we deem the work and thinking of scientific communities truly framework dependent, their trusted normative critics *must* come from outside the community. The initial source of destabilization must be external. Yet, as noted, a scientific field can only be transformed from within. It follows therefore that any historiography of scientific paradigm shifts faces two main tasks: first, to discern the science's external sources of trusted normative criticism, and then to explain how and by whom their subsequent challenge was converted into an internal transformative force.

Two of Michael Friedman's most interesting deviations from Kuhn will serve as our points of departure in what follows. The first, to which we first turn, has to do with the third level of discourse with which Friedman supplements Kuhn's two, namely, his misleadingly named notion of a "meta-framework" or "meta-paradigm," which serves in his account to "facilitate agreement or consensus across [a science's] different paradigms."[7] Friedman's second departure from Kuhn, with which we shall deal further down, is the place he assigns in the development of physics to such intermediary figures as Galileo, Helmholtz, and Poincaré.[8]

First let us turn to the question of a science's trusted critics. Some sciences face hostile opposition from religious and other ideological quarters. But such criticism will seldom be trusted in the desired sense, and is almost always dismissed offhand. Who, then, are a science's trusted normative critics, whose critical discourse is capable of prompting practitioners of voice and standing to reconsider its normative basics? In what sense can science be at all described as operating in a critical environment capable of the kind of transformative impact of which we have spoken? Who outside the boundaries of the community can be in a position to stimulate the sort of normative ambivalence toward key components of its prevailing scientific framework?

Friedman formulates the question of rationality very differently, of course, but his first deviation from Kuhn is nonetheless instructive. Because their work is *not* determined by the scientific paradigm in force, philosophers, mathematicians, engineers and even theologians, he argues, are capable on occasion of thinking the scientifically unthinkable without running up against heartfelt commitments. Such extrascientific discourses, he argues, provide science with valuable resources for novel and unconventional ideas and possibilities. But as noted previously, Friedman gives no account of the rational *incentive* a scientist could have for seriously entertaining such ideas. DiSalle, by contrast, suggests in Popperian fashion that the motivation to do so would be located in the critical stance certain of the science's practitioners adopt toward the framework constitutive of their work and thinking—scientists who in the course of "conceptual analysis" come to deem the constitutive framework of their science normatively wanting. But, as I have argued in the previous chapter, DiSalle gives no account of how adopting such a stance is at all possible without forsaking the framework's very constitutive force and function.

I fully agree with Friedman's point regarding the crucial importance to scientific rethinking of the discussion of scientific ideas in neighboring fields. And of course, I fully concede DiSalle's central point about identifying rational incentive with self-criticism. However, while they[9] insist on locating the initial source of criticism within the community, and the initial resources for meeting that criticism outside it, I propose the exact reverse: namely, to locate the potential sources of scientific questioning in such fields as philosophy, mathematics, and engineering, viewing them less as providers of *alternatives* to scientific thinking than as providers, by virtue of their potential ambivalating effect, of incentives to scientists for seeking for such alternatives themselves. To this end, with regard to both the extent and motivation for such professional extrascientific engagement, and its potential critical content, I find that Peter Galison's interactive notion of scientific "trading zones"[10] offers a richer and a more useful perspective than Friedman's more hierarchical imagery of meta-frameworks or meta-levels.[11]

PRACTITIONERS ABROAD

Like Friedman's *Dynamics of Reason*, Galison's interesting account of the give and take between a science's subdisciplines and neighboring cultures also builds on and away from an essentially Kuhnian, "antipositivist" picture of scientific development.[12] But, unlike Friedman's, it is not motivated by the problem of the rationality of paradigm shifts—at least not explicitly. Galison's objective is more squarely historiographical: a desire to get the story of science right by doing justice to the sense in which he believes physics as a whole exhibits a far greater stability and continuity across major changes in theory than Kuhn's picture would have us think.[13] He does not (or, at least, should not, since he does not make the point explicitly) oppose the idea of major Kuhnian paradigm shifts occurring *within* any one subculture of physics. The only Kuhnian idea he opposes explicitly is the assumption that physics as a whole "marches in lockstep with theory."[14] "It is now commonplace that the political dislocations of the French Revolution did not alter economics, social structure, politics and cultural life in the same measure," he notes wryly, and it is therefore "high time that we recognize that the physics community is no less complex."[15]

In other words still, Galison remains unmoved by the problem of incommensurability across framework transitions *within* each of the subcultures of physics, and subsequently—quite unlike Friedman's so-called meta-paradigms—his trading zones appear at least not to be meant to solve it.[16] The explanatory function of the trading zone, as he describes it in Galison 1997, is to account synchronically for the "social, material, and intellectual mortar binding together the disunified traditions of experimenting, theorizing, and instrument building" in physics,[17] and not, as in Friedman, to diachronically mediate "transitions between radically different [consecutive] conceptual frameworks."[18] Nonetheless, despite its apparent irrelevance to the questions that exercise Friedman's *Dynamics of Reason* and the present study, Galison's notion of a scientific trading zone proves extremely helpful in ways he himself failed to anticipate.

Indeed, at one point Galison's argument for the trading-zone approach reads like a direct attack on Friedman's approach of four years later. It is when Galison explains how his "original hope" to offer a "laminated description" of physics that would "underline the heterogeneity of practice within the wider physics community while allowing continuities on one level to bridge discontinuities on another" proved futile.[19] The diversities of belief and practice between the subdisciplines of physics, he argues, prove too great to reasonably claim that one community's stable paradigm could mediate or bridge the paradigm shifts of another. A discontinuity between the successive incommensurable paradigms of one community cannot be bridged by a third, incommensurable with both—even if it remains continuous and relatively stable across the shift. The relative unity of physics as a whole is achieved differently, he argues, not by "intercalating" the successive paradigms of one subculture by means of that of another, but by intercalating the subcultures themselves on the basis of their local interdependencies. Galison's key idea is to view the subcultures of physics as trading with each other in designated "trading zones."

> Like two cultures distinct but living near enough to trade, they can share some activities while diverging on many others. What is crucial is that in the local context of the trading zone, *despite* the differences in classification, significance, and standards of demonstration, the two groups can collaborate. . . . Anthropologists are familiar with different cultures encountering one another through trade, even when the significance of the objects traded—and of the trade itself—may be utterly different for the two sides. And with the anthropologists, it is crucial to note that nothing in the notion of trade presupposes some universal notion of a neutral currency. Quite the opposite, much of the interest of the category of trade is things can be coordinated . . . without reference to some external gauge.[20]

In his more recent work, Galison has come to acknowledge the fact (already pointed out by others)[21] that scientific trading normally extends far beyond the give and take conducted between practitioners of a science's "subcultures" on which he focuses exclusively in *Image and Logic.*

> We need this kind of approach not only to grasp the relation of algebraic geometers to string theorists—but also the way soil scientists speak with farmers, the way fishermen speak with fisheries regulators, the way research scientists speak with students. . . . In the heartland of fundamental physics, theorists in the 1970s worked hard to put their quantum field theories on the computer so they could calculate processes that eluded pencil and paper. It turned out that the very act of re-writing the theory into computer-readable form led to a *conceptual* shift in the way the physicists understood the very foundation of their theories; it led to a major shift in practice—and to a way of viewing theory, known as effective field theories, that to this day dominates everyday reasoning in these ethereal domains.[22]

And the list could and should be extended further to include all forms of professional scientific engagement and interfacing, such as "trading" for financial or public support, interacting with other scientific disciplines, offering scientific opinion in courts of law and scientific advice elsewhere, deliberating the didactics of how to best teach one's subject,[23] as well as engaging one's students, the media, the wider public, and the historians and philosophers of one's fields.[24] Trading zones abound.

From the perspective of the present study, Galison's metaphor is useful, first and foremost, for the way it brings to the fore the professional, interest-driven purposefulness of such interactions, which Friedman's structural notion of meta-levels lacks entirely. For although Friedman raises the problem of accounting for the way rational agents shift their allegiance rationally from one scientific framework to another like no one before him—which Galison does not—his proposed solution is not framed in terms of human agency at all, but focuses entirely, as I have shown in detail elsewhere, on the structural relation that successive scientific frameworks bear to one another.[25] Somewhere between framing his question and formulating his answer to it, Friedman's scientific philosophy ceases to be an account of scientific practice and deliberation, and morphs into a description of scientific *produce*. As a result, agency falls out of his picture of rationality altogether. Galison, by contrast, does not address the question of the dynamics, or the rationality of framework transition in science. And yet, by attending to the way trading is conducted between scientific cultures and their professional environments, his *Image and Logic*, more than any other work in the field, focuses on the discursive strategies employed by individual practitioners addressing one another across framework divides. Because his concerns are different from those of the present study, Galison, we shall see, remains oblivious to certain features of his model that will prove pertinent to ours.

Much of what transpires at the trading zone is unanticipated. Practitioners normally travel there to engage the representatives of adjacent fields with specific shopping lists in hand. As I argued against Friedman, central as they may be to their original fields, the mere *availability* of such offerings will not suffice to render them live scientific options for traders from other fields. Galison's novel proposal to view engaging practitioners from different fields as *trading* rather than as mere conversation partners, forcefully captures the essential self-critical element inherent in all rational endeavor. Professional traders frequent trading zones equipped with specific shopping lists that mark what they perceive they lack, and are constantly on the lookout for opportunities they failed to envisage. The trading they actually accomplish will often extend beyond their initial lists, as unanticipated offerings are brought to their attention. (Professional fairs are primarily geared to exposing potential buyers to new and unforeseen possibilities.) But even then, the dealing will seldom be idle. Unanticipated offerings will be "purchased" only when conceived as needed, that is to say, when conceived as new solutions to existing problems or as improvements to existing solutions. The very idea of agency in trading settings usefully presupposes an inherent measure of self-critical incentive. We shall return to this important feature of Galison's model in a moment.

Mediating Inter-framework Understanding

A second important aspect of Galison's metaphor concerns not their place, or motivation, but the *medium* by which such transactions are conducted. For members of different cultures and disciplines to meet and do business, more is needed than a designated "zone" and a desire to trade. In order to trade, they need to be able to converse with one another. The most novel aspect of Galison's account is his analysis of how scientific trading zones, like all trading zones, function by means of pidgins and creoles: mediating interlanguages that combine appropriately limited versions of the parties' professional vocabularies. For to coordinate their needs and expectations, they must be able to present their needs and offerings in a manner accessible to those with whom they trade.

Galison senses that the notion of a "mediating contact language"[26] for scientific trading can provide a philosophical corrective to the problems surrounding Kuhn's accounts of framework incommensurability and framework transition. But his remarks in passing to this effect miss the mark completely.

> Moving away from the stack periodization typical of conceptual schemes, radical translations, gestalt switches, and paradigm shifts comes at a price: we lose the vivid metaphorical imagery of totalistic transformations. In the place of such all-or-nothing mutations we need some guidance in thinking about the local configurations that are produced when two complex sociological and symbolic systems confront one another. Anthropologists are familiar with such exchanges, and one of the most interesting fields of investigation has been the anthropological linguistics of pidginization and creolization; I have used these ideas throughout the preceding chapters to characterize the border zones between different theoretical, experimental, and engineering cultures.[27]

As we shall see shortly, Galison is certainly right to assume that his notion of the scientific trading zone holds an important key to overcoming the problem of rational framework replacement—namely, that of not having to reduce it to an act of conversion or a gestalt switch. But not at all in the way he implies. In the brief passage quoted, something in his argument has become seriously undone. First, the problem of rational framework replacement is primarily *not* that of accounting for how two such frameworks "confront one another," but that of accounting for how the need for a new framework can be convincingly articulated out of, or from within, its predecessor—which comes down to that of accounting for the ability of practitioners of standing to adopt a critical stance toward the norms constitutive of their speech, practice, and thinking.

Second, as already noted, Galison's book is uninterested in, and nowhere even mentions, the *diachronic* problem of framework transition or modification. He introduces the notion of scientific pidgins and creoles as a way of accounting for the *synchronic* coordination achieved by traders across the boundaries of "different theoretical, experimental, and engineering cultures" and those of their other professional environments.[28] This has very little to do with the diachronic, with "the stack periodization . . . radical translations, gestalt switches, . . . paradigm shifts, [and] imagery of totalistic transformations" within the confines of any one paradigm-constituted scientific culture. *Image and Logic* is simply not about the dynamics of modifying or replacing single paradigms.

Finally, it is not at all clear what Galison means in describing scientific trading zones as places where traders from different cultures "confront one another." On his description, trading zones are where they come to trade, not to fight; the pidgins and creoles they develop are designed and employed to facilitate transaction, not dispute!

Nonetheless, the trading zone is a rich and fruitful metaphor that serves Galison well in crafting a history of modern microphysics that endorses what he believes is right about the "antipositivist" critique of positivism, while avoiding what he believes is wrong—namely, Kuhn's theory-driven, "crazy-quilt"[29] account of scientific revolutions as "block-periodized . . . 'all-inclusive' breaks"[30] within physics *as a whole*. As I noted earlier, even if Galison is granted his point, this is not an argument against Framework Dependency as such, only against the idea that physics as a whole is constituted by a single theoretical framework. I have no quarrel with that. The problem is that breaking the "lockstep" of a science's subcultures does not solve the problem of the rationality of framework modification and replacement for the framework dependency of each such subculture to itself.

Galison, as noted, does not at all address the problem of the *rationality* of such transitions. The problem that he implies is solved by the trading zone is the so-called problem of relativism, which he claims is solved not by breaking the lockstep of science as a whole, but by the ability afforded to scientists to talk across their framework barriers by means of the trading zones' pidgins and creoles. Viewing a science "as a densely connected map of distinct cultures bound by interlanguages,"[31] he argues in the introduction to *Image and Logic*, safeguards it from the "form of relativism" advocated late of Kuhn by Barry Barnes, David Bloor, and Harry Collins, which denies that "science [can] have any purchase on 'truth' or 'the world.'"[32] This is a curious statement indeed coming from someone who, in the closing paragraphs of the book, aligns himself philosophically with Davidson's and the later Putnam's decidedly framework-dependent accounts of truth and reality. The existence of interlanguages, developed to effectively facilitate the level of communication needed for trade and transfer of knowhow across cultural borders, can have no bearing on the traditional questions of truth and realism Galison seems to be hinting at in his opening chapter, for the simple reason that interlanguage cannot get those it serves *beyond* language. Intercalating *cultures* (by means of an interculture) and intercalating culture and reality are categorically different undertakings. Like Friedman (with respect to the diachronic axis), Galison too (across the synchronic axis) seems to think (at least in the opening chapter of *Image and Logic*) that by tuning down the radical abruptness of Kuhnian paradigm shifts, the challenge of "cultural relativism" can somehow be met.[33] But this is an illusion.

The problem the present study faces, however, is not that of truth or realism, but that of accounting for the rationality of framework replacement. And, on this count, in view of our discussion of it so far, Galison's trading zone would seem to have little to offer. It is certainly a step up from Friedman's meta-levels in terms of the self-critical attitude premised by its very notion of deliberated trading. But this type of self-critical awareness is precisely *not* the kind of potentially ambivalating normative criticism in which we are interested. Insofar as the felt deficiencies and needs that motivate practitioners to engage neighboring disciplines represent scientific concerns, they pertain at best to the type of self-questioning Carnap called "internal," namely, questions that arise for practitioners *within* their field, as opposed to questions that pertain to the failings and difficulties that are perceived to plague the normative framework constitutive of the field. In this sense, rather than generate ambivalence, trading with a view to resolving such difficulties, however pressing, will serve to reinforce commitment to the prevailing framework.

Exposure and Self-Questioning

Is there a sense in which practitioners can find themselves *challenged* in trading-zone settings to the point of destabilizing their allegiance to the framework constitutive of their thinking? As it stands, Galison's analysis offers little in this respect. As he sees it, the meeting ground facilitated by the trading zone's mediating pidgins and creoles is cordoned off from the deep ontological commitments that divide the trading parties, creating a safe space of shared phenomenology, as he puts it. But I believe he takes his metaphor a step too far, and in doing so misses something crucial about the world of interdisciplinary scientific "trading" and its likes.

Here is how Galison describes the "pidginized . . . foreigner version" of relativistic quantum field theory fashioned in 1964 by two prominent theorists, James Bjorken and Sidney Drell, as a textbook for experimentalists:

> Despite the radical difference in ontology—the set of what there is—a meeting ground exists between the experimenters and the quantum field theorists. That trading zone forms around the description of the phenomenological world of particle physics: How do photons recoil from electrons? How do electrons scatter from positrons? How do photons create pairs of electrons and positrons in the near presence of a proton? How are magnetic moments calculated for the muon? For these and similar questions, experimenters and theorists come to agreement about rules of representation, calculation, and local interpretation. Bjorken and Drell's volume 1 is an attempt to create a stable pidgin language that mediates between experimenter and theorist. Reduction of mathematical structure, suppression of exceptional cases, minimization of internal links between theoretical structures, simplified explanatory structure—*these are all ways that the theorists prepare their subjects for the exchange with their experimental colleagues.* I take these moves toward regularization to be the formal language analogues of phonetic, syntactical, and lexical reduction of natural languages.[34]

If Galison's notion of a scientific trading zone is to hold the key to normative transformation in science, we need to look less to what traders *bring to* the meeting ground—in terms of both their offerings and their shopping lists—and more to what they are liable to encounter there and *come away with.* Galison's rendering of the made-simple, "foreigners'" edition of Bjorken and Drell 1964 concentrates exclusively on the former. It emphasizes two important aspects of the mediating work of scientific interlanguages, but overlooks two others.

It lays great stress on the "watering down" involved in making one's science simple—the way "the fullness of theoretical representation would perforce be restricted (pidginized)";[35] how major practitioners like Bjorken and Drell bring to the trading zone significantly diluted versions of their worldviews, which not only render them easier to grasp, but present them in ways that play down their "radical differences" while emphasizing the phenomenology they are able to share. In this sense, a foreigners' pidginized version of a scientific language necessarily contains less than the full-blooded original it purports to make simple, especially with regard to their more technical and intricate dimensions, of which Galison speaks. Only the field's initiates possess the technical terminology, literacy, and know-how to grasp and appreciate their community's equivalent of "the fullness of theoretical representation" employed by the second, quantized theorists Galison describes.

But in one important respect, consistently overlooked by Galison, foreigner "pidgin" versions most often contain much *more* than the in-house, full-blooded originals they simplify. For when abroad, practitioners are frequently required to articulate in simple terms (to pidginize, if you wish) elements of their worldview that are so taken for granted in their home communities that they are passed over in silence in their professional in-house discourses, where they remain largely *unarticulated*. These are the things, Andy Warwick and Harry Collins have taught us, that seasoned researchers internalize to the point of *not* having to justify or explain in the course of their proofs and arguments—the things novices are taught "go without saying" in the field.[36] But when presenting to others, such basics often need to be made explicit—and the allusion to Brandom is not unintentional. (This is why even nonforeigner textbooks are generally bad examples of full-blooded professional languages, for textbooks do not normally represent the professional language as spoken by the "natives," but rather texts addressed to trainees and novices that are written in the pidgins and creoles of the community's initiating trading zones.)

Not all trading zones are the same in this respect, of course. Parties to interfaith dialogue, or peace negotiations, for instance, tend diplomatically to tread carefully so as not to offend their trading partners by unintentionally pushing too hard on matters of principle or faith. But with regard to science, such reservations are extremely rare. On the contrary, when representatives of different scientific cultures "trade" with nonscientists or with each other, more times than not they are met with real curiosity, even fascination, and subjected to the friendly prompting and interrogation of parties who are genuinely interested in what they do and how they think, and who jump at the opportunity of receiving an authoritative firsthand report.

Galison's oversight may be charged to the fact that in the "real" trading zones that figure in anthropological and sociological studies of "real" pidgins and creoles, the parties are usually far more cagy and reserved in this respect. Trading across ethnic or religious boundaries is often considered culturally threatening and potentially corrupting by at least one of the sides. But because trading is indispensable, the trading zone must be made safe by ensuring that negotiations are limited to what each side seeks to purchase from the other, while cordoning off the broader issues that sorely divide them. Trading is enabled and safeguarded by interlanguages that subtly serve the double function of both contact media and *buffer zones*. However, while scientific pidgins resemble "real" pidgins in the former sense, they hardly ever do so in the latter sense. Scientific traders engage each other with genuine interest, fascination, and, at times, even incredulity, not with suspicion or mistrust. Unlike "real" pidgins, scientific pidgins are neither designed nor employed with a view to *shielding* the communicating parties from one another's worldviews. Insofar as the parties entertain "radical differences in ontology," as in Galison's example, their meeting place will most often be characterized less by a polite and considerate *bracketing* of foundational commitment than by lively, good-natured, unthreatening curiosity as to how the other side sees the world. It is hard to conceive of anything like a scientific analog to the kind of fear and suspicion typical of "real" trading zones.

This is not to say that making explicit one's innermost professional intuitions and commitments is always an easy or a happy exercise. Many people have great difficulty articulating, let alone explaining, what they take for granted. Others find it less difficult than disconcerting. Either way, few, if any, in the scientific world would consider it threatening or potentially corrupting to be honestly prompted by fellow academics, technicians, university management, students, grant officials, or the media to openly explain the presuppositions of their worldview—as is often the case in "real" trading zones.

There is an important sense, then, in which scientific trading zones and their like offer opportunities with regard to scientific language of a nature quite different than those discussed by Galison. For it is only when "trading" with such significant others that practitioners are required to explicate, explain, discuss, and field questions and queries regarding the normative frameworks constitutive of their professional work. And it is here, when trading away from home, at the trading posts they frequent, far more than amid their professional milieus back home, that, I suggest (*pace* DiSalle, and going a large step beyond Galison), scientists are liable to be exposed to the kind of ambivalating normative criticism they and their peers are incapable of self-leveling. For it is here, far more than back home, that they are likely to encounter individuals who are sufficiently interested, yet sufficiently impartial, and therefore trusted, whose curious and probing promptings oblige them to articulate and justify that which in the course of their normal scientific work requires neither.

Often the very act of making explicit a taken-for-granted norm or constitutive presupposition, in response to such prompting—real or imagined[37]—is enough to create the self-estrangement required to render them ambivalent and consequently subject to self-criticism. Hearing oneself think aloud in response to the innocent probings of the uninitiated can suddenly and unexpectedly strike one as hollow and unconvincing. But more often than not practitioners prompted to explain themselves do so comfortably and confidently, without experiencing any sense of self-estrangement. In these cases, it is the incredulity of their "foreign" interlocutors or their very different way of thinking that may prove sufficiently "ambivalating" to set the transformative process in motion.[38] Be that as it may, the important point is that owing to the scientific trading zone's special discursive conditions, practitioners away from home are liable to be exposed there to the kind of destabilizing normative critique discussed in the previous chapter that is capable of rendering them normatively ambivalent.

The little that Friedman has to say about the initial phase of scientific framework transitions runs as follows. Mathematicians, philosophers, and other nonscientists are described as occasionally engaged, as part of their own in-house deliberations, in theoretically exploring options that fly in the face of the scientific sensibilities of their time, even to the point of absurdity. Mathematical work on non-Euclidean geometries during the nineteenth century is a good example. Scientists attentive to such discussions are pictured as rationally taking certain of these suggestions to be scientifically viable "live options," which they are liable, in turn, to incorporate in devising for their sciences a different constitutive framework from the one that presently informs their thinking. Friedman is right to insist that if rational framework transition is possible in science it cannot be an exclusively inside job. From the inside, from within the framework of classical physics, for instance, non-Euclidean geometry is scientifically unthinkable. But every other component of his picture of such transitions begs all the relevant questions. Because he is unable to describe the scientists responsible for seeking to replace their framework as having *reason* to do so, he ends up having to tell a story too close to Rorty's for comfort.

Why would scientists be paying attention to what mathematicians, philosophers, theologians and their like happen to be talking about in the first place? Friedman does not say, but as his account stands, he has only two options: pure chance, or Rortian irony. Talk of non-Euclidean geometry could have come to a physicist's attention by accident, or because, like a Rortian strong poet, he was constantly on the lookout for new and exotic opportunities that might prove useful in retrospect. In view of their blind contingency, neither option is very attractive as the first stage of a process purported to be rational. The more deliberately contrived, and hence potentially reasoned, of the two is the Rortian option. On a Rortian showing, it is rational to listen in on what mathematicians and philosophers have to say, because their musings might prove fruitful to science in ways unimaginable at present. In this sense, explains Friedman, though scientifically ludicrous, they are nonetheless "live options" for such scientists. But what could "live option" mean in this context? What could motivate a fully trained and fully committed, functioning classical astrophysicist to consider a non-Euclidean geometry "as both mathematically and physically possible (as, in Kantian terminology, *really* possible)"[39] to the extent that it "would have been irrational, unreasonable and irresponsible"—in their "own terms"—"to fail to consider it as a live alternative."[40] If all Friedman wants to say is that once such an option is on the table, it is reasonable for some such physicists to toy with it with the hope it may prove interesting, he remains firmly in the Rortian rubric! For such physicists to have *reason* to consider it a live option requires that they be at least arguably wary of the framework to which they are committed. But that is not what counts for a live option for Friedman!

And the same goes for the final stage. What could count as a reason for such physicists not only to consider the new, scientifically absurd geometry a "live option" *in theory*, but to invest serious and creative work in an attempt to replace the framework to which they are still committed, if deeming it scientifically wanting is not an option? In what sense could the new framework they are fashioning be hoped to prove the *better* of the two? What criterion of normative *preference* might be at work, if a framework cannot be critically evaluated from within? Again, Rortian irony seems to be Friedman's only real option!

Extending Galison's account of the scientific trading zone as suggested enables us to account for the initial, outward-facing stages of scientific framework transitions in a manner that begs none of these questions. Scientists do not just happen upon the work of people outside their fields or follow it out of mere curiosity, but professionally engage it as a matter of professional course. They frequent the trading zone not aimlessly, but for a variety of specific and purposeful reasons: seeking solutions to particular problems that defy the tools in hand (conceptual, mathematical, technical), support for their projects (financial, institutional, political), new recruits to their fields, and so forth. And while they are there, they will be on the lookout for new and unanticipated offerings, just as Friedman describes. This is what *brings them to* the trading zone: a purposefulness of intent and openness that is wholly framework determined and therefore fully and unproblematically rational. Options chosen at this stage are deemed to be "live" exactly because they answer, in ways anticipated or not, to well-formed, framework-determined difficulties and needs.

Kuhn's Misleading Imagery

The foregoing is all perfectly rational, but does not, of itself, have the power to rationally motivate framework replacement, as Friedman's problematic account implies. Normative modification and realignment can be undertaken rationally only when motivated by normative self-faulting, and normative self-faulting, as we have seen, necessarily depends on the ambiguating effect of trusted external normative criticism. The *fact* that possibilities that run afoul of a practitioner's scientific norms and standards are considered by nonscientists to lie outside his field is, of itself, no more potentially destabilizing than the existence of any other unschooled opinion. Mere *awareness* of normative diversity—that in other communities people think differently than oneself—seldom constitutes enough of a challenge to one's own system of norms to render one normatively ambivalent in any sense.[41] When it does happen, it is not because one is presented with a different viewpoint, so much as when one's own viewpoint is interrogated and questioned! And this, I have argued *pace* Galison, is what practitioners very often encounter at the trading zone.

The most misleading of Kuhn's metaphors is his likening of paradigm shifts to acts of conversion. Bracketing the social and communal aspects of converting, a convert is a person who loses his faith in one framework or form of life *in favor of another*. Someone who only loses his faith, or who has developed doubts or misgivings about it but is still in search of an alternative, is not a convert. One converts *to*, not away from. Dissatisfaction and parting with the old is a necessary condition of conversion, but what *makes* for conversion is the irresistible pull of the new. This is why conversion is the paradigm of an unreasoned shift, like falling in love. But the point I wish to make is different, and does not concern reasoning. The point is that without saying so explicitly, using the metaphor of conversion to describe scientific framework transitions firmly premises a well-formed alternative framework being in place before the shift in question. One can only convert to a framework that already exists. And the same goes for the image of a gestalt switch. From a gestalt point of view, one can only switch from one perceptive form-generating whole to another. And there is no sense in which either of them can be deemed problematic, even from the perspective of the other! In other words, Kuhn's imagery strongly implies an exclusive focus of attention on what I have called the second stage of the transition—that in which practitioners are faced with the problem of choosing between the framework they are committed to and the well-formed alternative in the offing.

The same goes for Kuhn's much discussed notion of incommensurability, which pervades his account and the discussion to which it gave rise to a far greater extent than the metaphors of conversion and gestalt. For in both its semantic and normative senses,[42] "incommensurability" serves as an essentially backward-looking assessment of the semantic and normative relations between successive scientific frameworks, once again necessarily presupposing that a fully developed version of the new is in the offing.

The problem is that Kuhn's vocabulary has profoundly informed the discussion not only among his followers, but also among the opposition. Whether one is arguing for or against Kuhn's contention that successive scientific frameworks are sufficiently incommensurable to prohibit rational discourse between them, the question will always be framed as a second-stage query, as a problem that arises only after a new framework is in place. In this way, one could say, Kuhn's imagery and vocabulary do more than focus disproportionally on the second stage; they work actively to eliminate the first stage from the narrative structure of paradigm shifts. The creative and deliberative process of framing and structuring the new framework, awakening to the need for one, gradually articulating what can be and needs to be replaced, what has to be retained, and how to somehow combine them is, by virtue of Kuhn's very language, written out of his account of scientific revolutions, to the point of denying it any part in their structure.

Hence, for someone like Friedman, determined to prove Kuhn wrong, at least with respect to mathematical physics, the problem of the rationality of framework transitions in science remains framed in terms of overcoming incommensurability, even when the question he is addressing is a decidedly first-stage issue. The only way Friedman can envisage the rational formation of a new scientific framework is to place it in a second-stage setting, that is, to describe its main elements as *already existing* in professional discourses conducted beyond the confines of the scientific community in question, which are nonetheless susceptible to its members' scrutiny as live, and hence commensurable, options. The problem for Friedman, and for all who have written in response to Kuhn, is never the problem of finding one's own framework wanting, but that of being able to understand and appreciate a framework radically different from one's own. And for that reason, the problem is not framed and engaged by Friedman as one of agency, but as one to do with the structure and partial nestedness of successive scientific frameworks—that is to say, as one to do, as noted above, with the scientific *product* rather than with its practice.

In Galison's account of the trading zone, incommensurability is (supposedly) overcome by the lowering of barriers and bracketing of differences enabled by the employment of mediating interlanguages. But overcoming the problem of incommensurability only proves that practitioners of the old framework were capable of understanding and considering certain radically diverse available options. It is powerless to explain *why* they endorsed them, and in what sense it was rational for them to do so. However, when it is realized, as argued here, that the trading zone enables more than a display of normatively diverse *options*, that its pidgins and creoles provide the discursive setting for practitioners to engage in pointed, deep-reaching, and, hence, potentially ambivalating normative *questioning* of the kind they are never liable to experience in their home communities, the road is open to analyze framework transition in science as a problem of agency.

THE TRANSFORMATIVE FORCE OF CREATIVE INDECISION

The creative individuals initially responsible for rationally setting in motion the transformation of a field are thus to be sought among those who were lucky to be exposed to the ambivalating challenge of real or imaginary trusted external critics. But that is not enough. Normative ambivalence is a necessary, but not a sufficient, condition for framework transition. Individuals successfully rendered ambivalent to what Haugeland calls their "existential commitments"[43] may end up thinking differently from their colleagues, or even considering a career change, but that of itself will not suffice to transform their *field*. To have the required impact on the normative commitments of their home communities, it is not enough for certain practitioners to have become normatively ambivalent. Their ambivalence must somehow carry, spread within the community, and similarly ambivalate a critical mass of fellow researchers. We have seen (in theory) how normative ambivalence might be produced; we now turn to examine how it might be disseminated and consumed.

In keeping with the conclusions of the previous chapters, the argument has so far proceeded from the outside in, as it were, with the arrow of explanation pointing *from* social setting *to* personal deliberation, in an attempt to characterize the kind of intersubjective discursive setting most conducive to destabilizing intrasubjective normative scientific commitment. We now need to reverse the arrow of explanation in order to account for the impact of ambivalated individuals fresh back from the trading zones on their home communities. Trading zones are only frequented by individuals. Whole communities do not travel abroad, so they can only be ambivalated at home, from within, by already ambivalated members of sufficient standing and voice to be heard and adhered to. If there is any truth to my argument so far, then the role of such individuals in the history of major scientific upheavals is decisive. But their story has largely yet to be told. Current historiography of science—diverse, wide-ranging, and richly reflective as it has become in recent decades—passes such figures over in near silence. Historians of science show little awareness of their existence and hardly any interest in exposing and analyzing the sources of their ambivalence and its effect on their peers. Attempting to tell their story raises two related questions: that of their visibility, and that of the nature of their impact.

Out of Sight

First, to visibility: Ambivalent, undecided scientists of the past are hard to see. For reasons mentioned above, Kuhn's account of paradigm shifts unwittingly locates the entire drama of framework transition around its decisive, dramatic end point, after the new framework has been framed and developed, and the community is required to choose. It makes its case by juxtaposing vivid "before" and "after" pictures in striking, seemingly unbridgeable contrast, as in a weight-loss advertisement. It is easy to see how this kind of tilted narratology remains virtually blind to the intermediary figures of the preceding stage, struggling with foundational doubts and uncertainty, whose commitment to the old is waning with still no solid alternative in sight.[44] And as I explained in chapter 1 with reference to my earlier work, those whose historiography of science leans more toward Popper than to Kuhn are no better off. There, too, those responsible for the intermediary moments of deliberative *indecision* are likewise lost from sight in the blinding dazzle of the knockdown *experimentum crucis*, the decisive refutation, and the bold and novel conjecture. The idea, central to the account proposed here, that the driving force of normative reassessment should be located primarily in moments of discordant ambivalence and painful indecision sits uneasily with virtually all current historiography of science. Ambiguity and indecision are, as a rule, not considered the stuff of science, at least not the stuff capable of contributing to understanding the drama of bold scientific advancement.[45] The interesting exception is Friedman, who, despite his Kuhnian focus on the second stage of the process, emphasizes the role of such intermediary figures, if for the wrong reasons.

But first, it is important to understand that the invisibility of ambivalent, undecided scientists owes to more than historiographical blinkering. It is not as if such individuals could be clearly seen had historians not been looking the other way. They are obscured from the sight even of their own communities, for the simple reason that doubts and indecision are hardly ever considered worthy of *publication* even by those whom they plague. Modern science is a predominantly written culture whose worldview is deliberated, cultivated, enriched, modified, and proliferated within the community primarily by means of research papers. And the scientific research paper is not where one normally doubts and dithers. Doubters and ditherers as a rule do not view their ambivalence and indecision as worthy of scientific publication, and nor do their potential editors, referees, and readers. Indecision and doubt are seldom regarded as contributions to knowledge, unlike the kind of reasoned suggestions for new developments and reasoned challenges to old developments that are deemed publishable. It is true that current historiography of science remains largely uninterested in the ambivalent and undecided members of the communities it studies, but it is equally true that the ambivalent and the undecided remain largely hidden from scientific view within their communities, where their work is relegated almost entirely to the private realm of the diary, the journal, the logbook, the personal memoir, and personal correspondence—none of which are capable of carrying their master's voice very effectively into the public, professional domain of scientific discourse proper.

Fossils of Indecision

Nonetheless, undecided, ambivalated practitioners of standing occasionally do succeed (if somewhat unwittingly) in making their indecision known to and felt by their peers in ways capable of effectively ambivalating the field. Friedman's historiography of modern mathematical physics is an interesting first case in point.

As noted, the way Friedman frames the problem of the rationality of scientific framework transitions is deeply influenced by Kuhn. For him, the obstacle to be overcome is not that of normative self-criticism, but that of bridging the supposed incommensurability of successive frameworks. To do so, Friedman elects to redescribe paradigm shifts as smoother, less ruptured, "evolutionary" (to use his term) rather than revolutionary processes of transition, in which new frameworks are developed naturally and gradually out of their predecessors—which they continue to preserve as special cases. "Practitioners of succeeding paradigms," he writes,

> are not helpfully viewed as members of radically disconnected speech communities . . . On the contrary, successive paradigms emerge precisely from one another, as succeeding stages in a common tradition of cultural change. In this sense they are better viewed as different evolutionary stages of a single language rather than as entirely separate and disconnected languages.[46]

In Friedman's decidedly more leveled picture of paradigm shifts, figures whose work occupies halfway positions between the two end points of the transition become important for the way their work furnishes the "essential intermediate stage[s]" between the old and the new frameworks that render the latter "actually continuous with" the former.[47] Two of his main examples are Galileo's treatment of free fall and projectile motion, which occupied a halfway position in the transition from Aristotelian to Newtonian mechanics, and Poincaré's work on the foundations of geometry, which occupied an analogous position in the transition from classical physics to relativity theory, "without which," Friedman declares, "it is hard to imagine how the use of non-Euclidean geometry in physics could have ever been envisioned as a real possibility."[48] Friedman describes neither work as ambivalent or undecided, or in any other way unsure or troubled. On the contrary, he presents them as confident attempts to retain certain elements of the old, while developing certain novel departures that would eventually form the basis for the new. Such mediating, intermediate positions, he argues, provide living disproof of the radical unbridgeability of Kuhn's depiction of the transitions they so nicely straddled.

Friedman is right in objecting to the idea of meaning incommensurability and in not describing the mature works of Galileo and Poincaré as anything but confident presentations of positions to which, by the time they were published, their authors were fully committed. Tycho Brahe's hybrid grafting of a heliocentric planetary system on a basically geocentric cosmology would be another example of a midway position presented in genuine confidence and with full conviction. But accepting this does not commit one to Friedman's account of the *role* of such intermediaries in the course of framework transition, or to what he believes the positions they held represent. The intermediary frameworks devised by Tycho, Galileo, and Poincaré, I insist, were no more "natural developments" of those they purported to replace than were the final positions of Kepler, Newton, and Einstein. That they were able to form and maintain such halfway positions only proves that even they, on whose work the old framework evidently still exerted a profound hold, had succeeded in becoming sufficiently ambivalent toward it to propose and endorse several radical departures. The relative "nearness" of the positions they formed to both those they abandoned and those later adopted by their respective communities does not absolve Friedman of the need to account for their *reasons* for forming them.

According to Friedman, the key elements of the new framework take form elsewhere, and are gradually introduced into science ready-made, *not* as means to replace or modify a scientific framework deemed *faulty*, but as a series of natural *extensions* of a well-functioning framework perceived by its practitioners to be "uncontroversial and unproblematic."[49] He continues to pay lip service to Kuhn's picture of revolutionary scientific development, but his rhetoric sounds less and less convincing as his argument develops, bordering more and more on the oxymoron. Viewed prospectively, "from a properly historical point of view," he writes toward the end of part 1,

> we see . . . that, although there is indeed an important sense in which succeeding paradigms are incommensurable or non-intertranslatable, the concepts and principles of later paradigms still evolve continuously, by a series of natural transformations from those of earlier ones. . . . we can thus view the evolution of succeeding paradigms or frameworks as a convergent series, as it were, in which we successively refine our constitutive principles in the direction of greater generality and adequacy.[50]

He gives no account of what he means by "greater adequacy" in this context. Had he done so, he might well have found ways to locate the rationality of framework replacement in the felt *inadequacies* of the old. But he leaves the phrase philosophically unaccounted for. And thus, the most unsettled and exciting moments of creative scientific innovation are redescribed by Friedman in extraordinarily placid terms that give no sense at all of *why* those who made them saw fit to do so. "When we move from the Aristotelian framework to that of classical physics," he writes, as if ticking a checklist,

> We retain Euclidean geometry intact, discard the hierarchical and teleological organized spherical universe, and modify the Aristotelian theory of natural motion—in such a way that we retain the idea, in particular,

> that there is a fundamental state of natural motion following privileged paths of the underlying geometry.

Likewise,

> When we move from classical physics to special relativity, we again retain Euclidean geometry, and also, of course, the law of inertia, but we now move from a three-plus-one dimensional to an essentially four dimensional spatio-temporal structure—which, however, yields the older three-plus-one dimensional structure as a limiting case.[51]

And the same goes for the subsequent move to general relativity.[52]

Adamant to present such changes as rational, yet unable to describe them as attempts to *rectify* a framework deemed wanting,[53] Friedman is forced retell them as attempts not to *overturn*, but only to naturally extend, expand, and generalize the framework they purport to replace. According to Friedman, the transitions in question were not only perceived as natural, but were relatively smooth and modular—hence the important role he attributes to intermediary figures such as Galileo and Poincaré. But to describe even Galileo's halfway theory as a natural expansion and generalization of Aristotelian physics and cosmology is like describing Copernican heliocentrism as an expansion or generalization of Ptolemaic geocentrism! Almost every one of the new elements of Galileo's system squarely contradicts the Aristotelian element it replaced.[54]

More important in the present context is that, almost by definition, such halfway attempts—and, again, Galileo's is a good example—exhibit, in addition to profound breaks with the systems they replace, severe internal tensions. Unable to make a clean break with the past, far from representing a first, natural step away from former convictions, such works attest to a strained split, or dithering, between the serious commitment of their authors to the frameworks constitutive of their earlier work and a deeply felt need to modify those same frameworks. Such works bear vivid witness, I insist, not to how easy and natural, but to how *difficult* it must have been for their authors to forge and maintain such departures from their initial commitments. Their historical and historiographical importance lies much less in how far they ventured away from the old and how closely they approximated the new, than in the tense and jarring juxtapositioning of old and new they insisted on maintaining.

Such are also the curiously hybrid, early Victorian meta-scientific positions briefly described in chapter 1: George Peacock's strained twofold account of algebra; William Rowan Hamilton's insistent talk of two, rather than one, "science[s] of dynamics"; and William Whewell's better thought-out, yet similarly strained "antithetical" theory of scientific endeavor and accomplishment. The likes of Tycho, Galileo, Peacock, and Poincaré all presented their hybrid proposals with confidence and conviction, but I insist, contrary to Friedman, that there was nothing continuous or natural about their radical and painful breaking with the party-line geocentrism, Aristotelianism, mathematical realism, and Kantianism from which they had distanced themselves, respectively. I prefer to see such fundamental departures as the highly creative work of deeply ambivalated practitioners, whose anxious and novel compromises mark how far, when forced, they were capable of going.

Such works were indispensable to the deeply transformative moments in which they partook, as well as to the latter-day historian, because of the way their Solomonic attempts to split their subject matter in the hybrid manner described dramatically captured and displayed their normative ambivalence. Carefully analyzed, pried apart at their (usually) rough seams, and properly reenacted, such doubles and splits offer historians much toward recapitulating the destabilized mindset of their authors. But for that they must be read and analyzed *prospectively*, the way history is lived and should be studied, as acts of anxious, urgent, and imaginative engagement, rather than retrospectively, as mere midway stations between one stable, consistent framework and the next. They should not be written off, as they so often are, as unimaginative first tries, or as the work of muddled reactionary rear-guarding. Properly read, they present historians with what might be described as fossilized normative dilemmas that preserve for posterity frozen, ossified snapshots of the tortured yet highly creative indecision of their ambivalated authors that forced other leading practitioners to take a stand.

Indeed, as noted, normative doubts or ambivalence do not spread easily as such within well-functioning scientific disciplines. But confident resolutions of such doubts can. And most often the first attempts to resolve such doubts are proposed by the "ambivalated" individuals themselves. For who else but them would be motivated to do so? In Frankfurtian terms, normative ambivalence amounts to being of two minds with respect to certain elements of one's normative framework—a form of indecisive dithering with regard to those elements. To the remainder of their framework—namely, the part of their I-self to which they are not ambivalent—such individuals remain wholeheartedly committed. It follows that even when their ambivalence is resolved by replacing the doubtful elements by others, the remainder of their framework will remain firmly intact. In other words, the kind of hybrid, halfway solutions proposed by Tycho, Galileo, Peacock, Whewell, and Poincaré are exactly what one would expect in the first instance!

Modest as their suggestions might seem in retrospect, few scientists are capable of such creative feats of rethinking and self-dividing, however deeply ambivalated they become. It takes a Tycho Brahe to replace a fully fledged, mathematically functioning, Ptolemaic, geocentric cosmology with a heliocentric planetary system embodied within an otherwise geocentric cosmology; a Galileo to replace a two-tiered, hierarchical universe that displays for each of its two parts a different kind of natural motion, with a unitary, nonhierarchical universe that incorporates both forms of natural motion in its one part; a Peacock to propose a systematic account of modern algebra as a hybrid juxtaposition of "arithmetical algebra" that "gives rise by suggestion" to a purely "symbolical algebra," sufficiently general to accommodate its full range of algebraic expression; and a Poincaré to replace Kant's account of the epistemic exclusivity and necessity of Euclidean geometry with an equally constitutive notion of the same "form of intuition," but adopted by convention. And, again, there was nothing "natural" or "naturally continuous" in any of these startlingly original proposals! On the contrary—and this is my main point—such hybrids do *not* seem natural and continuous to the fellow practitioners to whom they were originally addressed, but rather strike them as forced and jarringly disjointed. Once the coherent unity of the old system has been broken and key elements of an alternative system have been put into play by practitioners of such standing, other members of the community are liable to be rendered ambivalent with respect to larger portions of the framework, and motivated to develop the transition much further, as was the case with regard to all four of those discussed here.

INTERLUDE

The Story So Far and That to Come

Following on from the autobiographical account related in chapter 1, the previous three chapters have argued that the idea formative of the discussion of scientific development since Kuhn, that it is possible to commit oneself to robust notions of scientific rationality as self-criticism or Framework Dependency, but not both, must be readdressed.

The argument set forth from a view of rationality shared in dialogue with such thinkers as Sellars, Brandom, and Korsgaard—a view of rational agency that deviates sharply from much of the current literature on rationality, especially of the game-theoretic variety. The difference between defining rational action as acting for a reason and as action that gets what one wants, prefers, or needs might seem slight, but it is not. Only the former views agency as necessarily definitive of the rational. To act for a reason requires reflection and judgment, while a move that proves beneficial to maximizing an actor's utility does not. As understood here, rationality in the former sense necessarily turns on first-person prospective assessment; rationality in the latter sense, by contrast, is, in principle, a matter of third-person retrospective evaluation.[1]

The argument proceeded as follows: To act for a reason requires one to *have reasons* for acting (as opposed to impulsive incentives one does not consider reasons), and to *find* reason to *act* (as opposed to remaining inactive). Both selecting one's reasons and deciding to act upon them, it was argued, involve a definitively central element of normative self-criticism. In other words, taking such a view of rationality necessarily commits one to Critical Rationalism—if of a variety that steers significantly closer to latter-day appropriations of Kant and Hegel than what usually goes by that name.

Yet at the same time, a look at the work of such writers on agency and personal identity as Korsgaard, Taylor, and Frankfurt revealed that normative self-criticism of the kind premised in both practical and theoretical reasoning is inherently and necessarily framework dependent. In order to keep one's world and oneself in normative check and hold oneself accountable, a system of norms must already be in place. In other words, taking such a view of rationality just as necessarily commits one to Framework Dependency!

Critical rationalists have overlooked this point entirely. Popper himself dismisses the "myth of the framework" altogether, implying (at best) that in his opinion the only norm involved in critical rationalism is the universal and wholly transcultural standard of formal consistency. And Bartley, if subtly, follows suit. I have argued that the idea that scientific research can be thought to proceed, or even ideally proceed, with no criteria of scientific propriety other than logical consistency is not only ridiculous but inconsistent with Popper's own extensions of critical rationalism to the politics of the Open Society and to social engineering. Others, like Agassi and Berkson, and even Lakatos, have taken a far more reasonable stand in recognizing the formative role played by "hard core," meta-scientific commitment in determining the agenda of first-order scientific enquiry. However, even they maintain, unjustly, as I have argued, that instrumental considerations alone suffice to undermine this kind of normative commitment. What kind of instrumental reasoning can dislodge firm commitment to mathematical Platonism or an instrumentalist view of quantum mechanics or a realist view of particle physics, for example? How can a normative framework fail adequately to interpret a practice of which it is constitutive, as Agassi believes it can?

If rationality commits one to Critical Rationalism, and Critical Rationalism to Framework Dependency, there are only two ways to avoid having to contend with the problem of normative self-criticism—both, as I have argued, quite unacceptable: (i) to deny that scientific framework transition *can be* rational, that is to say, undertaken for a reason, as implied by some of Kuhn's early pronouncements; or (ii) to insist that the merits and demerits of scientific frameworks are rationally gauged against a fixed meta-framework of objective, transcultural, thin scientific ideals, such as accuracy, precision, scope, simplicity, fruitfulness, and consistency, as claimed, for instance, in some of Kuhn's later work.[2]

This, then, in very brief summary, is the philosophical motivation for the present study's insistence on contending with the problem of accounting for the rationality of framework transitions in science by showing how it is possible to remain faithful to Critical Rationalism even when it is applied reflectively to the framework it presupposes in criticism, yet without diluting that framework or the seriousness of the commitment it commands. It is possible, I have argued, but not within the traditional, deliberative, intrasubjective confines of Critical Rationalism. These, we have seen, need to be transcended twice over.

First, even our keenest and most rigorous self-critique cannot get us beyond the norms we exercise in so criticizing. And the more earnest our normative self-questioning, the more profound our commitment to the norms our very questioning employs. It is not as if calling our own norms into (normative) question is difficult to do, because it requires a keener form of standing back from ourselves. It is impossible to do, because it requires that we stand back *devoid* of self! In a sense this is the unproblematic flip side of what G. A. Cohen called the "paradoxes of conviction."[3] Normative commitment pertains to a deeper form of acculturated internalization than reasoned acceptance,[4] to the extent that awareness of the fact that our deepest convictions are the contingent product of the way we were raised and trained is incapable of dislodging them. (Cohen's own autobiographical examples nicely liken his professional training [in philosophy at Oxford] to his devoutly communist upbringing as a child.) Cohen considers paradoxical the fact that normative commitment remains unaffected by such knowledge.[5] Building on Frankfurt and others, this study considers it an unproblematic aspect of the primacy in principle of normative commitment to all forms of reasoning—which is just another way of saying that left to our own devices, we are incapable of normatively faulting the norms to which we are committed.

However—and this is the philosophical point that grounds the approach argued for here—exposure (at times, even imagined exposure) to the normative critique of others enables one to transcend the limits of purely intrasubjective self-assessment. This is what I have loosely dubbed, late of, but very differently from, Brandom, "the Hegelian" aspect of the discussion. But it is not the *force* of our critics' arguments that counts, because arguing against our heartfelt norms, however forcefully, can never convince. What counts is that our critics be *trusted*—not to argue *validly*, for that we have seen is impossible, but to argue *truthfully*. For although powerless to persuade us, when deemed *sincere*, such criticism can convey a clear sense of how our normative commitments appear to them. And this, I have argued, can be sufficiently destabilizing to render the norms in question self-criticizable. In other words, to appreciate the transformative impact that our trusted critics may have on our steadfast normative commitments requires deviating even further—some might say too far—from what is customarily regarded as rational exchange. Such criticism cannot convince, but it can on occasion ambivalate sufficiently to prompt serious reconsideration. Normative *self*-criticism is possible, but only in the context of a wider, intersubjective, normatively diverse critical discourse.

Lacking such an account of normative self-appraisal, historians of science committed as was Kuhn to the framework dependency of scientific practice and produce have been disinclined to analyze cases of framework transition as the products of rational human endeavor. There has been a marked tendency in post-Kuhnian science studies to tell the story of such transitions by describing them in collectivist and/or institutional terms, or, as in Friedman's case, by subtly reassigning the term "rational" from the *process* to the *product* of scientific reasoning—as if a *descriptive* account of the changing *produce* of scientific endeavor will suffice to yield a *normative* assessment of the endeavor itself.[6] Telling their story the other way around, as the reasoned and contrived products of a process of self-critical deliberation, I have argued, minimally requires breaking it down into three basic and ordered components, the first and third of which pertain to the process, and the second to the creative product of the kind of normative ambivalating described above. The first component will focus on how practitioners of standing were rendered ambivalent toward central components of the prevailing framework in the course of "trading" away from their home communities. In the second, the novel and widely read accounts of their field produced as a result will be shown to have been profoundly and uneasily split along the fault lines of their initial indecision, and analyzed accordingly. The third component will show how these works then succeeded in prompting other practitioners of standing to take a more decisive position.

As I noted in chapter 1 with respect to my own historical work, the order of discovery of the three components will usually not follow their chronological order. Direct evidence of the first and third stages will normally lie hidden from sight deep within the pages of private correspondence, intimate diary entries, and early draftings of later work. What will inevitably first attract attention are the hybrid published works that characterize the second stage. If they are noticed and duly appreciated (their confident rhetoric notwithstanding) as works of "creative indecision" (rather than mere confusion), then their origin and formation can be traced back through the archives to the ambivalating circumstances that begot them. Likewise, once thus appreciated, their careers thereafter can be traced forward through the impact they had on the work and thinking of the key practitioners who read and responded to them.

So much for theory. Part 3 of this work, to which we now turn, applies the narratological model for which I have argued to one major aspect of the four case studies briefly outlined in the opening chapter. Going back to some of the thinkers I have worked on in the past, in the light of the philosophical and historiographical insights to which that work eventually gave rise, allows me to come full circle in the manner indicated at the outset of this study.

I would have liked the following chapters to work first back and then forward from the hybrid meta-scientific texts produced respectively by Peacock, Herschel, Whewell, and Hamilton: to tell the full story of their crucial roles in the framework transitions to which they were party; to read them as creatively symptomatic of profound and ultimately unresolved meta-scientific ambivalence, tracing them first back to the ambivalating circumstances (the particular trading zones) responsible for their authors' indecision, and then forward to the impact they had had on the thought of others. But that is impossible within the confines of the present study. As indicated at the outset, each of them owed to a somewhat different set of ambivalating circumstances and pertained to quite different forms of meta-scientific unease. And yet, the stories of their formation and impact intertwined and interacted to weave a picture larger and more significant than the sum of its four individual components.

To do full justice to each of the four micro-historical stories of intrasubjective deliberation, and then to show how they combine to tell a broader story of the critical middle years of a major meta-scientific transition, will require at least a separate monograph. Part 3 of the present study will instead exemplify my thesis by focusing in some detail on one of the four, namely, that of Peacock's algebraic work. I have chosen to focus here on Peacock because the story of the making and impact of his *Treatise on Algebra* bears on the other three stories and intertwines with them at crucial junctures. The same could be said of the making, though perhaps less of the impact, of Whewell's historical and philosophical work. But I prefer to focus on Peacock for two main reasons. First, I have already devoted a book-length study to the emergence of Whewell's historiography and philosophy of science, to which I remain by and large committed. Focusing on Whewell would have required casting the story I told there in the terms of the present study, of course, and especially highlighting the hybrid and ambivalent nature of Whewell's published work, of which I had only recently become more fully aware. Still, the larger contours of the story, as well as most of its details, would have remained much the same.

Second, and perhaps more important, we are far less accustomed to speaking of framework transitions in mathematics than in science or philosophy. Attending closely to Peacock's case affords us a close and fine-grained look at the workings of a major meta-mathematical realignment within mathematics proper, and at least a glimpse the process it set in motion with regard to mathematical physics.

The three chapters that make up part 3 tell Peacock's story along the lines developed in chapter 4. Chapter 5 describes the background to the young Peacock's mathematical career—his involvement in the Analytical Society as an undergraduate at Trinity, alongside fellow students John Herschel, Charles Babbage, and a little later, William Whewell, and the role it played in revolutionizing the university's mathematics curriculum—and the keen ambivalating challenge to his native "Lockean" view of algebra he encountered especially in Babbage's work on "pure analysis."

Chapter 6 describes Peacock's impressive attempts during the 1820s to address the challenge, first in a book-length study of the history of arithmetic in which he sought to establish algebra as generalized arithmetic on a richer footing than was customary. It was a failure, but as I shall argue here for the first time, a telling one. This led, in the second half of the decade, to the formulation of his inherently hybrid *Treatise on Algebra*, published in 1830, which is analyzed in some detail as the product and expression of profound meta-mathematical ambivalence.

Chapter 7 then describes how four major figures responded to the *Treatise*: Augustus DeMorgan, William Rowan Hamilton, the young and short-lived D. F. Gregory, and Peacock's Trinity colleague, William Whewell. All four, as we shall see, treated it as a major contribution to be seriously reckoned with, and all to some extent misread it in the light of their own perspectives. And all four responded differently in ways that combined to herald a new and essentially formalistic era in British mathematics, much earlier than in Germany or France. I end the chapter by describing the transformation of Peacock's own thinking as a result, as is evident in the second edition of his work, published fifteen years later and too late to be effective. I leave it to others to assess the depth and lasting worth of the framework transition set in motion by Peacock's *Treatise*, and present it here as a case study of the kind of creative indecision that, as this work argues, is the key to understanding how framework transitions can be rational.

PART 3

They (a History)

CHAPTER FIVE

Peacock, Babbage, and the "Heresy of the Dots"

INTRODUCTION

The strained, two-sided meta-scientific positions, briefly outlined in chapter 1, that Herschel, Peacock, and Whewell each came to occupy during the 1830s were significantly different, yet owed, I believe, to a common point of departure: their involvement during the 1810s and early 1820s in the reform of Cambridge mathematics. Herschel and Peacock came up to Cambridge in 1809 and were joined by Charles Babbage a year later. Together they founded the Analytical Society, most probably in the autumn of 1811. This was a student society devoted to "re-importing" the now "exotic" achievements of eighteenth-century continental analysis to Cambridge after what they described as "nearly a century of foreign improvement, and to render it once more indigenous among us."[1] It was a bold and timely venture on all counts,[2] but one that would prove in the long run considerably unsettling for all involved, except perhaps Babbage. Herschel, Babbage, and Peacock, who became close and lasting friends, were responsible for all the publications related to the society. Whewell, who came up in 1812, joined them a year later. I devote this chapter to looking again at the story of the short-lived Analytical Society and its immediate aftermath[3] with a view to reconstructing the initial setting in and against which Peacock's conundrum would take form alongside those of Herschel and Whewell.[4]

The reform of Cambridge mathematics has attracted considerable attention among historians of early nineteenth-century British science, but its meta-scientific significance has yet to be properly appreciated. In virtually all of the studies related to the life and work of Herschel, Peacock, Babbage, and Whewell, their early involvement in the Analytical Society and its campaign is noted, but mostly as marking where and when the special bond between them was initially forged, before embarking on their more significant scientific endeavors. It is deemed formative of their lasting friendship, and of the first phase of their respective careers, but, almost without exception, not of their subsequent thinking.

This is an unfortunate oversight, if to some extent understandable. In view of the rise of mathematical physics in England thereafter, the reform process set in motion by the three twenty-year-olds is justly considered a crucially important and fitting venture, in which more prominently placed individuals had previously failed.[5] Yet at the same time, it is conceived as a more or less straightforward matter of making up for lost time, rather than of breaking new ground—as an almost technical corrective leveling, which, although it sounded strange and "exotic" to Cambridge-trained ears, did not represent a framework transition in the sense discussed in previous chapters. The society is at best described as campaigning for the replacement of Newton's dot notation by Leibniz's *d* notation, and of the "synthetic," geometrical methods associated with Newtonian fluxions by the "analytical," algebraic methods of Euler, D'Alembert, Lagrange, and Laplace.[6] In both respects, historians have followed the actors themselves, who at the time described their campaign exactly thus—both as preaching the obvious and as campaigning for a *d*-style algebraization of the calculus. The Analytical Society ceased to exist as such by the late 1810s, and its mission of reform was well on its way to completion by the early 1820s, but it would take Herschel, Peacock, and Whewell the best part of an additional decade to seriously realize and begin to face up to what they had gotten themselves into. By then the reforms they had campaigned for had long been successfully instituted, and were seemingly of no further consequence.

It is this second, deeper, and more reflective phase that holds the key to the significance of the Analytical Society's original undertaking to the story I wish to tell. Its significance lies not in the replacement in general of geometry by algebra as grounds for the calculus, and certainly not in issues of notation per se, but in the young rebels' opting specifically for Lagrange's version of the calculus. To do so was to adopt an approach that was not only thoroughly *algebraic*, but thoroughly *formalistic*, which led, I shall argue further, to disagreement among the young rebels as to the nature of algebra itself.[7] The exchange of dot for *d*, and of geometrical methods for analytical (that is, algebraic) ones, as Grabiner and Guicciardini, and to an extent even Playfair long before them, argue, could be, and in fact were easily endorsed by Newtonians during the eighteenth century, because they represented little more than acts of *translation* of the notation and method they replaced.[8] But by adopting a formalist approach to the calculus, and certainly to algebra at large, the young rebels unwittingly introduced a major framework transition, to which they awoke only much later and would grapple with (or in Herschel's case, consciously abstain from grappling with) throughout the 1830s and beyond.[9] Failing to appreciate the meta-scientific significance of their initial Lagrangianism (as did they themselves at the time), current historiography, even among historians of mathematics proper, has consequently failed to take note of their eventual awakening to it. Studies such as Guicciardini's history of fluxions;[10] Enros's, Becher's, and Schweber's accounts of the Analytical Society;[11] Koppelman's, Pycior's, and Richards's work on early Victorian algebra;[12] and most of the existing biographical studies of Herschel, Babbage, Whewell, DeMorgan, Hamilton, Gregory, Thomson, and Maxwell all but ignore this second, longer-term, reflective phase of the reform process. And thus, in the larger scheme of telling the story of their illustrious careers, the early involvement of Herschel, Babbage, Peacock, and Whewell in the Analytical Society's reforms is remembered more as a fortunate early crossing of paths, than as a richly significant moment that proved crucially and disturbingly formative of their thinking thereafter.

The Philosophical Breakfast Club

The belated ambivalating effect of the society's Lagrangianism owed to the fact, of which the three friends only later became aware, that it jarred painfully with other of their heartfelt commitments. This second set of more native convictions is the subject of Laura Snyder's biographical study of the intertwined lives and works of Herschel, Babbage, Whewell, and political economist Richard Jones. Peacock is unfortunately left out of her story. Snyder's book, entitled *The Philosophical Breakfast Club*, is a significant contribution even though it ignores the mathematical context almost entirely and would, therefore, seem to offer little to the project at hand.[13] But *The Philosophical Breakfast Club* differs significantly from all earlier studies in insisting that the intellectual involvement of their student days was centrally and crucially formative, not only of their lifelong friendship, but of their lifelong understanding of, and engagement in and with, science.

Snyder argues rightly that as undergraduates, Herschel, Babbage, Whewell, and Jones shared more than a commitment to continental analysis. They were just as deeply committed to a Baconian vision of science, to which, she argues, they remained faithful throughout their careers. And as with continental analysis, their commitment to Bacon also translated into reformative action. Working back from a letter written by Thomas Forster to Whewell in 1841, fondly recalling "the period when you and I and Herschel and Babbage used to meet at our Sunday morning philosophical breakfasts,"[14] and from their later researches and writings, Snyder reconstructs what she dubs "the Philosophical Breakfast Club"—the group of undergraduates, among them Herschel, Babbage, and Whewell, who were also deeply involved in the Analytical Society, and Jones, who was not.[15] This group, she claims, would meet regularly in Herschel's rooms at St. John's for Sunday breakfasts after chapel to discuss Bacon's vision and methodology of science and to debate how it should be best understood and implemented. "The four men," she claims,

> came together in agreeing that Bacon's exhortations had gone unheeded: science, [they] thought, was not in much better shape than it had been before Bacon's day. . . . Babbage, Herschel, Jones and Whewell knew that science should be reformed, and they pledged themselves to the task. The Philosophical Breakfast Club would develop an up-to-date version of Bacon's method, and spread the word—in books, articles and speeches. They would disseminate the image of the scientist as a social reformer, and show how scientific discoveries could be used to help society . . . And they would remake the institutions of science to be more like "Solomon's House" in *New Atlantis*. They would, as Babbage put it to Herschel, "do their best to leave the world wiser than they found it."[16]
>
> What is most amazing about this breathtakingly ambitious program is that these four men did bring about a revolution in science greater than any Bacon had hoped to spark in his own day.[17]

Some historians may be wary of Snyder's colorful description of the Breakfast Club's meetings. Some may well contest there ever being such a "club"—the name is hers—for there exists scant direct historical evidence of their meetings other than the letter to Whewell, mentioned above, written almost thirty years after the event. Writing in 1864, Babbage, by contrast, describes his "chief and choicest" of several undergraduate associations as a "breakfast club" of his own, at Peterhouse, consisting "of some ten or a dozen friends who usually breakfasted with me every Sunday after chapel; arriving at about nine, and remaining between twelve and one o'clock. We discussed all knowable and many unknowable things." He makes no mention of Bacon.[18]

But even if Snyder's account of the Philosophical Breakfast Club is deemed more a narrative device than historical fact, it is both powerful and compelling. The young Herschel, Babbage, Jones, and Whewell's native commitment to Bacon's vision and philosophy of science was deep and real, and centrally formative of their scientific work and thinking thereafter.[19] The circumstantial evidence may not conclusively corroborate Snyder's vivid description of the Sunday gatherings at St. John's, but it certainly corroborates the main thesis of her book regarding the role played by Bacon's philosophy in the four friends' subsequent thinking and its early roots in their student days. In the 1859 "Prefatory Notice" to his edition of Jones's unpublished papers—a text also referred to by Snyder[20]—Whewell writes warmly of Jones's laudable inductivism and its early roots:

> Having noticed the inductive nature of Mr. Jones's social and political philosophy as its special and distinctive character, perhaps I may be allowed to say that the disposition to take such a course in his speculations belonged to him from an early period. It existed at the time of his Cambridge undergraduateship, and was nourished by the sympathy of some of the companions of his college days. The *Novum Organon* was one of their favorite subjects of discussion. John Herschel was one of the companions to whom I have referred; and his incomparable *Discourse on the Study of Natural Philosophy* shows how admirably such previous trains of thought had fitted him to be the commentator of Francis Bacon. And, if I may speak of myself, I may mention that, when I dedicated to him[21] my *History of the Inductive Sciences*, I said, in the dedication, that my volumes "were the result of trains of thought which had often been the subject of our conversation, and of which the origin went back to the period of our early companionship at the University," recollecting with pleasure and gratitude that I had, even then, discoursed with them on such subjects. Richard Jones himself was always prompt and strenuous in maintaining that all the best part of his intellectual habits had been acquired at College.[22]

Treating the mathematical reforms of the Analytical Society as they are commonly viewed, as important, timely, yet essentially unproblematic correctives, the story told by Snyder of how Herschel, Babbage, Jones, and Whewell eventually "transformed science and changed the world" is driven by a philosophical incentive grounded exclusively in their Baconian point of departure.[23] Snyder is right in insisting that in the course of the four men's "undergraduateship" Bacon's philosophy was central, not only as the objective, but as the implement of their determination to reform science. But I differ with her account of the later phases of their careers—especially Whewell's. At no point does her story become that of an informed *critical assessment* of Bacon's view of science by any of her four heroes. It remains throughout the story of a shared, longstanding, and essentially unquestioned *obligation* to it. As a result, it is made to focus largely on the four friends' lifelong involvement in, and discussion of, the *empirical* aspects of scientific research, a story in which issues of observation, inductive inference, calculation, mapping, and precision figure prominently, to the exclusion of the kind of questions that Bacon largely ignored—namely, those to do with scientific theorizing, conceptualization, mathematization, and historical emergence, on which at least Whewell would write extensively.[24]

Snyder's narratology would seem to ring true more of the unfolding of Herschel's thinking with respect to science than of Whewell's philosophical development. Herschel's *Preliminary Discourse* indeed steered wide of questions of conceptualization, mathematization, and epistemology, holding to a philosophical course remarkably close to Bacon's original philosophical agenda. Whewell's philosophical work did not. He, too, set out as an avid follower of Bacon, but ended up, in his *Philosophy of the Inductive Sciences*, a keen critic of both Herschel and Bacon.[25] (Babbage and Jones, who did not venture seriously into the philosophy or historiography of science, are less interesting in this regard.)

Even with respect to Herschel, I doubt the story is as simple and straightforward as Snyder's book implies. Since the publication of Bacon's *Novum Organon* in 1620, much that was of philosophical significance had transpired in both science and philosophy. Sophisticated and timely as it had been in its day, Bacon's inductivist methodology was powerless to account viably for the sort of highly mathematized, richly theoretical physics that had swiftly developed in the wake of Newton's great work. Nor could it be viably revived from a philosophical point of view without giving serious consideration to at least Hume's empiricist critique of causation and the radical idealist alternative proposed by Kant. The great difference between Herschel and Whewell in both these regards is that the former's persistent Baconianism was only made possible by knowingly *ignoring* them, while the latter's eventual turn away from Bacon was prompted, in contrast, by taking them in all seriousness. (Notice how, in his 1859 "Prefatory Notice," cited above, Whewell described Herschel's *Discourse*, not as offering an authentic, philosophically viable, latter-day form of Bacon*ism*, but as showing him to be "the *commentator* of Francis Bacon," and how he described his own views, not as a commentary on or an interpretation of Bacon, but as the culmination of "trains of thought" that *originated* in their undergraduate "discourse" of Bacon—more than implying that they had culminated elsewhere.)

However within the narrow confines of the present study, I wish to remain focused on what I consider to be the uncontestable and crucially important contribution of Snyder's book: the way it firmly locates and vividly exposes the young reformers' early commitment to, and detailed engagement with, Bacon's philosophy of science. For it is against and in relation to this deeply set conviction that the part omitted from her story—their insistent opting for Lagrange—acquired for them its meaning and eventually unsettling significance.

Finally, it is an omission that also accounts for the different *dramatis personae* of the differently focused stories told by Snyder and myself. Peacock, who throughout his career showed no interest in engaging in, and only moderate interest in promoting, empirical research proper,[26] is written out of Snyder's story entirely, but owing to his deep involvement in rethinking the nature of mathematics, plays a major role in mine.[27] Richard Jones, by contrast, who contributed decisively to the bottom-up, inductive study of political economy, but who lacked any interest in mathematics (or in the philosophy of science, for that matter), plays a major role in Snyder's story, but plays virtually no role in mine.

Yet for all our differences in emphasis and interpretation, I am in agreement with, and indebted to, Snyder's insightful and forceful portrayal of the four friends' philosophical mindset during and far beyond the years of their involvement in the mathematical reforms to which I now turn.

RIPE FOR REFORM

"By the end of the eighteenth century," notes Warwick, "mathematics dominated undergraduate studies [at Cambridge] to the almost complete exclusion of all other subjects, a situation that continued until the mid-nineteenth century."[28] At the Cambridge to which the young Herschel, Babbage, Peacock, and Whewell came up to obtain their degrees, university honors and college fellowships were granted solely on the basis of mathematical competence. Until 1822 mathematical ability alone determined a student's ranking in the Tripos, the university's honors examination, and even after the establishment of the Classics Tripos in 1822, the right to compete in classics still depended on first obtaining honors in mathematics.[29] And yet, despite the unique position it commanded in the university curriculum and the vigor with which it was studied, the mathematics taught at Cambridge well into the 1820s lagged a full century behind continental developments.[30] The reason for both its prominence and backwardness was one and the same: the overwhelming legacy of the university's greatest son, Sir

Isaac Newton.[31] This, along with the Cambridge colleges' peculiar method of academic inbreeding, virtually sealed off the university from the swift process of "degeometrization" of the calculus[32] in the hands of continental mathematicians, the analytical turn introduced and led by Euler from the 1740s onward. "D'Alembert, Lagrange, and Laplace," writes Guicciardini,

> all worked along Eulerian lines. The object of their calculi were classes of single and multivariate functions, rather than curves or surfaces; partial differential equations and the calculus of variations were amongst the most important research areas of their production; analytic mechanics based on extremal principles provided a stimulus for, and was an application of, their algorithms. Moreover, they stressed the algebraic character of the differential and integral calculus, obtained solutions in terms of manipulations of symbols rather than in terms of geometric properties, and defended the need to separate the calculus from geometry.[33]

Staunchly committed to Newtonian fluxions and to its "pricked letter," dot notation (as opposed to Leibniz's *d* notation), Cambridge mathematics not only lagged behind, but remained hopelessly out of touch with, the new "Eulerian calculus"—the term used by Guicciardini to describe the analytical turn introduced in mid-century by Euler.[34] By the time the young Herschel, Peacock, and Babbage arrived at the university, even the keenest Cambridge-trained students were unaware of the existence of partial derivatives, for example, and had no means of articulating them if they had wanted to.[35] Enmeshed, as Boyer puts it, in "a web of provincialism,"[36] Cambridge was so far behind that by the time the three young men met up in 1810–11 and took gloomy stock of the mathematics in which they were expected to excel, only a revolution could rouse it from what Walter Cannon termed its "Newtonian slumbers."[37]

After decades of near-stagnation, and in an institution in which the study of mathematics was so highly valued, it was only a matter of time before a pressure group campaigning to reform the university's mathematics curriculum would take form. Given the Cambridge system, it is also predictable that in order to succeed such pressure would have had to gather momentum within the university, yet, at least initially, beyond the confines of the existing faculty. It had to begin among the students, but could only be effectively realized by members of faculty who had the authority to change the examination system. The so-called analytical revolution indeed started out as a student rebellion and later morphed into a "revolution of the dons," to use Sheldon Rothblatt's well-known title,[38] as former associates of the Analytical Society, first Peacock and Richard Gwatkin, and then Whewell, graduated and went on to obtain college fellowships.

Herschel and Babbage—The Revolutionary Duo

A critical mass of talented, discontented undergraduates wholly dedicated to the cause was needed to set the process in motion. But what were the chances of a young student raised and schooled in England even becoming aware that the mathematics he was earnestly taught at Cambridge was any different from, let alone seriously inferior to, the mathematics taught and practiced elsewhere? English schools did not teach any continental mathematics at the time, and they reinforced the university mathematics curriculum by viewing Cambridge studentships as something for which they were committed to prepare their best students, not to question. Fortunately, unlike the vast majority of their Cambridge peers, including Peacock and Whewell, both Herschel and Babbage had the privilege of very different, yet less conventional mathematics educations before arriving at their respective colleges.

Herschel, at age eight, was removed from Eton by his mother because he was bullied by older boys, and was sent to a private school in nearby Hitcham run by a family friend, Dr. Gretton.[39] According to Buttmann, Hitcham offered "a purely classical course," which his parents decided to supplement by employing a private tutor to introduce him to natural science, mathematics, modern languages, literature, and music.[40] Luckily for Herschel, the person hired for the job was the Scottish accountant and mathematician Alexander Rogers, whose patron was no other than the most eloquent and outspoken critic of English mathematics, the Edinburgh-based John Playfair![41] It is no coincidence that the closing section of Playfair's review of Laplace's *Mécanique céleste* reads like the Analytical Society's manifesto.[42] Playfair's fingerprints are all over Rogers's pedagogy—drawing Herschel's attention in 1808 to Silvestre Lacroix's work on the differential and integral calculus (presumably to the shortened, one-volume, textbook edition of 1802 which the Analytical Society would adapt and partly translate several years later),[43] and prompting him a year later to read Lagrange's *Mécanique analytique* as "good preparation" for the study of Laplace.[44] Personally coached by a close follower of Playfair, whose review of Laplace he would also almost certainly have read before leaving for St. John's College, Herschel could not have come up to Cambridge better prepared for the reformative task he would undertake.

Yet, although probably better *informed* of continental developments and their superiority than Babbage would be when he arrived at the university a year later, Herschel had apparently not shown much interest in mathematics, or mathematical promise, under Rogers's tutoring.[45] He would eventually shine, of course, but only a year or two later. Babbage's was a different story.[46]

Due to ill health his early schooling was, as one biographer describes it, "haphazard" and "chequered."[47] The several schools he attended and the two tutors he was assigned—one a Cambridge man, the other an Oxford man—offered very little by way of mathematical training. The young Babbage was wholly self-taught in this respect, making up for the lack of formal guidance with bountiful energy, endless curiosity, considerable talent, and hardheaded youthful zeal, all of which continue to vibrate between the lines describing that early "passage from his life" written fifty-four years later. Babbage simply loved mathematics! "Being passionately fond of algebra," he reminisces in 1864,

> I now employed all my leisure in studying such mathematical works as accident brought to my knowledge. Amongst them were Humphrey Ditton's "Fluxions,"[48] of which I could make nothing; Madame Agnesi's "Analytical Institutions,"[49] from which I acquired some knowledge; Woodhouse's "Principles of Analytical Calculation,"[50] from which I learned the notation of Leibniz; and Lagrange's "Théorie des Fonctions."[51] I possessed also the Fluxions of MacLaurin[52] and of Simpson.[53]

He thus arrived at Cambridge, as he puts it, with equal fluency "in the dots of Newton, the d's of Leibniz, [and] the dashes of Lagrange."[54] Lacking any systematic preparation for the mathematics he was about to be taught, yet with eclectic exposure and a passion for the subject, the eager young Babbage arrived at the gates of Trinity College in the fall of 1810, looking "forward with intense delight to the certainty of having [the "many difficulties" with which "I had . . . met"] all removed."[55]

But his euphoria was short-lived. His tutor and lecturers dismissed his "difficulties" as irrelevant to his course of mathematical studies at the university, because they "would not be asked at the Senate House," and firmly advised him to remain focused on the set curriculum.[56] Their cautioning had the opposite effect, of course, and Babbage found himself again following an unruly and unstructured course of mathematical reading of his own devising, "devour[ing] the papers of Euler and other mathematicians scattered through the innumerable volumes of the academies of Petersburgh, Berlin, and Paris, which the libraries I had recourse to contained."[57]

By the beginning of his second year, the "distaste" he had acquired "for the routine of the studies of the place,"[58] took an uncharacteristically more ordered and systematic turn, with the purchase of "the great work of Lacroix" at a London bookseller's on his way back to Cambridge in October 1811, upon which he immediately set out in earnest.[59] The purchase was neither incidental nor impulsive. Babbage paid a fortune for the book, which he evidently deemed to be absolutely crucial for his mathematical advancement.[60] What could have motivated him to do so? Babbage's late recollections in *Passages* are not always reliable,[61] but the idea conveyed by the MS 13 account—that he had purchased the book a year earlier, at the commencement of his first year—is downright ridiculous. What incentive would a schoolboy with no proper mathematical training and with sporadic knowledge of a dappled and accidental assembly of works, arriving at the university for the first time eager to be mathematically enlightened, have to chase down and pay such a fee for such a work before attending his first Cambridge tutorial? How would he have even known of its existence? Unless, of course, he was convinced to consult it by someone sufficiently aware of the sorry state of English mathematics and well enough versed in the mathematics taught elsewhere to strongly recommend the work—someone, that is, like Herschel's Mr. Rogers. But no one resembling Rogers was evenly remotely involved in Babbage's pre-Cambridge schooling. There is simply nothing to indicate[62] that he might have seriously considered correctively supplementing his Cambridge course of study before embarking on it! On the contrary, according to the more reliable of his two autobiographical accounts, he arrived at Cambridge delighted at the prospect of having his difficulties dispelled, only to have his high hopes dashed by the end of his first year. He had still not encountered a Mr. Rogers, but would certainly have befriended a well-trained Rogers *trainee*: the second-year star, Herschel![63]

At some point between arriving bright-eyed at the university for the first time in autumn 1810 and sitting down, Lacroix freshly in hand, "at the commencement of the second year of my residence at Cambridge," to famously parody the Bible Society poster, calling for a society dedicated to "the propagation of D's" and consigning "to perdition all who supported the heresy of the dots,"[64] Babbage's growing estrangement from the Cambridge system underwent a turn that I doubt could have been exclusively of his own making. Sometime in the course of that year, his frustration with teachers determined to limit his instruction to the Senate House questionnaire developed into an angry awareness of the marked discrepancies between the mathematics he was being taught and the achievements of continental analysis. By the end of his first year, he had come to realize that the problem was not his teachers' unwillingness or inability to respond to his questions, but the kind of mathematics the university insisted that they teach him. Paradoxically, *because* his tutors chose to make light of his difficulties and queries, he would have had no way of knowing whether the differences in notation and style his personal, "chequered" library exhibited were at all significant—his apparent personal attraction to functional equations notwithstanding.

If there was anyone capable of having that kind of effect on Babbage, it was Herschel—who, when the two first met, might not yet have exhibited the flair and desire for analysis he would do a little later. But he certainly would have understood and sympathized with the young Babbage's frustrations. More important, he would, in all probability, have given them context and depth by introducing him both to Playfair's review and to his copy of Lacroix. This would explain Babbage's sudden appetite for the publications of the St. Petersburg, Berlin, and Paris academies, and his determination to purchase and master Lacroix's more comprehensive three-volume masterpiece. It also stands to reason that Babbage reciprocated by infecting Herschel with his passion for algebra, enthusiasm for the "new math" he was swiftly mastering, and, of course, his growing interest in the calculus of functions. With Babbage's exasperation and mathematical creativity, now properly channeled by their newfound friendship, and Herschel's duly awakened enthusiasm, the formation in autumn 1811 of the dissenting and anti-establishment Analytical Society was only natural. The charismatic duo's combination of talent, angry frustration, "youthful and romantic ardor," and firm resolution "to leave the world wiser than they found it"[65] worked its magic in attracting the small group of fellow students and recent graduates capable of providing Herschel, Babbage, and the newly recruited Peacock with the social and semi-institutional setting for the work they needed to do.

But the Herschel-Babbage "double team" brought to the revolutionary moment toward which they were swiftly heading something more than youthful ardor and an appreciation of continental analysis, something closer to the concerns and main focus of the present study, something that all the better established, would-be reformers who came before them, as well as many of their friends who joined them in forming the society, lacked—most conspicuously Peacock. Herschel and Babbage brought with them a *clean fresh start*. The upshot of Herschel's early tutorship under Rogers and Babbage's eclectic self-instruction was that their commitment to continental analysis was primordial, rather than a later, learned realization that was adopted against a background of earlier convictions (as it was in Peacock's case, as we shall see). And as the spirited flurry of their foundational contributions to the calculus of functions[66] clearly attests, it was a commitment that premised a decidedly formalistic grounding of mathematics. For them, the initial choice of Lagrange's version of the calculus (rather than the limit-based approaches of Euler, D'Alembert or, later, Cauchy) was natural and wholly uncomplicated.[67]

THE ANALYTICAL SOCIETY

The first and only publication with which the Analytical Society[68] was associated by name was the "volume of our transactions," as Babbage described it,[69] or *Memoirs of the Analytical Society*, composed entirely by Herschel and Babbage, and published in 1813. The volume consisted of a lengthy, jointly authored preface[70] and three memoirs, or research papers, one by Babbage on continued products, and two by Herschel, on trigonometrical series and equations of differences, respectively. (Both of the authors had yet to turn twenty-two!) Judging by that alone (and the tendentiously research-favored description of the MS 13 account), one might wrongly conclude that the society was founded primarily as a mathematical society.[71] However, its founders, especially Babbage, Herschel, and Peacock, envisaged it as more than a forum for enthusiasts of continental analysis to meet, present their work, and compare notes. Their declared mission was to bring the mathematics taught and practiced at Cambridge in line with the state-of-the-art continental analysis they were practicing: "to re-import the exotic, with nearly a century of foreign improvement, and to render it once more indigenous among us," as Herschel and Babbage wrote in their introduction.[72] To this end, the young rebels knew full well that it would not be enough for learned mathematical research papers to be read and even published out of Cambridge. The university's curriculum, and most of all its examination system, would have to be revised. And for that to happen, suitable textbooks had first to be made available.[73]

Hints of Disagreement

Babbage, whose idea it was to form such a society, proposed to do so with a particular textbook project in mind: the partial translation into English of

Lacroix's 1802 *Trait élmentaire,* which he undertook jointly with Herschel and Peacock. There exists, however, a curious discrepancy between Babbage's latter-day recollection of his plan and the manner in which it had in fact been realized forty-eight years earlier. The oft-quoted passage of his 1864 autobiography, mentioned briefly above, is worth quoting more fully:

> At this period, Cambridge was agitated by a fierce controversy. Societies had been formed for printing and circulating the Bible. One party proposed to circulate it with notes, in order to make it intelligible; whilst the other scornfully rejected all explanations of the word of God as profane attempts to mend that which was perfect.
>
> The walls of the town were placarded with broadsides, and posters were sent from house to house. One of the latter form of advertisement was lying upon my table . . . Taking up the paper and looking through it, I thought it, from its exaggerated tone, a good subject for parody.
>
> I then drew up the sketch of a society to be instituted for translating the small work of Lacroix on the Differential and Integral <Calculus>. It proposed that we should have periodical meetings for the propagation of D's; and consigned to perdition all who supported the heresy of Dots. It maintained that the work of Lacroix was perfect [and] that any comment was unnecessary.[74]

Lacroix's *Trait élmentaire* did indeed provide a lucid introduction to the calculus, wholly cast in Leibniz's *d* notation. In this respect, it was unarguably perfect. The English edition, however, was not ready for publication until 1816. The cause of the delay was Peacock who, quite contrary to Babbage's witty parody, had insisted that the English-language version of the book be circulated among its English-speaking readers annotated by over one hundred additional pages of copious notes. Peacock apparently found the famous French textbook less perfect than Babbage's clever quip suggests that he, Babbage, had believed at the time.[75] Whatever Peacock found wanting in the Frenchman's presentation of the calculus, and he evidently found much, seems not to have impressed Babbage, at least in retrospect—if, of course, his autobiography can be trusted on this point.[76]

Personal, motivated memoirs written late in life and long after the event are bound to be distorted to some extent by age, hindsight, and a latter-day desire to settle scores. This is certainly true, as I have noted more than once, of Babbage's MS 13. Still, even that long after the event, knowing full well what the society's rendition of Lacroix ended up looking like, it is hard to imagine how Babbage could have unwittingly confused the parties to the

Bible Society dispute to the extent of imagining that he had parodied the wrong poster. It makes better sense, I think, to assume that, despite the time difference, the disparity between Babbage's recollection of his motives and plan for the society, and the textbook eventually produced on Peacock's insistence, attests to real disagreement. But disagreement between whom, when, and especially about what? As I shall try explain in the remainder of the present chapter and the first part of the next, I believe the disagreement was fundamental—that it went to the heart of the society's project, and subsequently to that of the deep-seated normative ambivalence responsible for the hybridic nature of Peacock's later work.

His disagreement with Babbage, I shall argue, developed in stages as his understanding of the mathematical enterprise took and changed form in tense dialogue with the radical alternative Babbage was formulating at the time. Peacock's position set forth from the agreed and shared Lagrangian agenda of the society, but it developed differently from that of his two comrades-in-arms. For Babbage, the society's program reflected the view of mathematics already manifest in his unfolding treatment of functional equations—a view he developed into a novel, bold and close to fully fledged "philosophy of analysis" by 1820 or so, but which would remain, alas, in manuscript form. Babbage's mature views, in other words, represented a direct and uncomplicated extrapolation of his earlier commitments, to which he would remain deeply committed throughout his life.

Peacock's trajectory was exceedingly more dialectical and troubled, and owed to a growing awakening to the extent to which the society's agenda, as understood by Herschel and Babbage, jarred, rather than dovetailed, with his prior convictions. Cast in the idiom of part 2 of this study, Peacock's views developed quite differently from Babbage's, as a direct result of the challenge to his initial commitments posed by the society's agenda, which was rendered joltingly explicit, I shall argue, in his "tradings" with Babbage regarding the nature of algebra. Peacock devoted the lion's share of the 1820s to attempts to bridge the ambivalating divide by arguing first "historically" (as in "natural history"), and then more systematically that the two jarring conceptions of algebra could be viewed as stages in a progression. The outcome constituted one of the most impressive, most influential (though not in the usual sense of the term), and at the same time, one of the most impossibly hybrid, accounts of algebra ever to be formulated—all of which will be taken up in detail in the next chapter. I shall devote the remainder of present chapter to the two elements that make up the immediate normative context in which Peacock's two-sided project was to take form: first, to clarifying more precisely what I mean by the society's agenda, and then to presenting and analyzing Babbage's important and understudied "Philosophy of Analysis" project.

Opting for Lagrange

Clever as they were, Babbage's now famous mock-theological quips about the society dedicating itself to "the propagation of D's" and to the "Principle of pure D-ism in opposition to the Dot-age of the University" should not be taken literally. As noted at the outset of this chapter, it would be seriously wrong to assume that the society aimed merely at notational reform—as if the group believed that Newton's dots were all that stood between continental analysis and Cambridge men. It would also be mistaken to assume that what was at stake was the difference between the *d* notation and the dots' *original* mathematical semantics—as if their aim was to replace Newton's original version of the calculus (which was still taught at Cambridge) with that of Leibniz. The society was not formed with a view to harking back to the ways in which the calculus was *originally* conceived. Their campaign concerned its present state; their aim, to replace the Newtonian calculus taught at the university with one particular current brand of what Guicciardini terms "Eulerian calculus"—the version proposed in 1797 by Lagrange in a work tellingly entitled: *Théorie des fonctions analytiques contenant les principes du calcul différentiel dégagés de toute considération d'infiniment petits, d'évanouissans, de limites et de fluxions, et réduits à l'analyse algébrique des quantités finies.*[77]

Newton's calculus endeavored to calculate the rate of change ("fluxion") of a "flowing" quantity changing in time ("fluent") by considering the limit values of what he called their prime and ultimate ratios. Leibniz's infinitesimal calculus aspired to achieve the same by considering the ratio between the infinitesimal differences (differentials) between successive values of the variables involved. However, supplying a mathematically viable account, geometrical, logical, or algebraic, of the notion of limit proved to be no easier than elucidating the notion of differential. Lagrange's calculus exempted itself from any such task by grounding the calculus in a formal, operational, rather than conceptual, definition of a given function's derivatives. The idea was as simple as it was radical. Brooke Taylor had famously proved that any continuous function $f(x)$ could be expressed as a power series by means of a given point x_0 in its domain, as follows:

$$f(x) = f(x_0) + (x - x_0)\, f'(x_0) + (x - x_0)^2 f''(x_0) \,/\, 2! + (x - x_0)^3 f'''(x_0) \,/\, 3! + \ldots$$
$$(x - x_0)^n \cdot f^n(x_0) \,/\, n! \ldots$$

Lagrange reversed the reasoning by proving the theorem without resort to the calculus, showing (he thought) that every well-defined function can be expressed by means of a power series around a given point x_0 as follows:

$$f(x) = f(x_0) + A(x - x_0) + B(x - x_0)^2 + C(x - x_0)^3 + \ldots$$

where the coefficients A, B, C, and so forth are determinable functions of x_0. Lagrange then formally *defined* the first derivative of $f(x)$ at x_0, $f'(x_0)$, as A, its second derivative, $f''(x_0)$ as $B \cdot 2!$, the third, $f'''(x_0)$ as $C \cdot 3!$, and so forth. He then went on easily to show how, thus defined, the function's "derived functions" conform to all the known laws of derivation, and yield the known derivatives of the various function types.

Unlike any other version of the calculus, Lagrange's derived functions do not, of themselves, mark or measure momentary rates of change, inclinations, ratios of differentials, or limits. They pertain to no inherent conceptual basis, and are, therefore, of themselves, as meaningless as the rules of grammar, or those of a game. The laws they obey follow from formal definitions and are, therefore, true *of* nothing. They may be *applied* to calculate or represent rates of change, instantaneous velocities, or limits, but not because anything inherent *to* them justifies such applications a priori. They are successfully applicable to such realms only because they prove so in retrospect.

To the best of my knowledge, Lagrange's version of the calculus is the first branch of mathematics to be founded as a thoroughly formal system in this respect. It also dovetails nicely with the stark "instrumentalism" of his analytic mechanics, as the position would come to be known a century later in the wake of Pierre Duhem's philosophy of science.[78]

Although the young Herschel, Babbage, and Peacock all fully conceded Lagrange's power-series version of the calculus, the disagreement that emerged at the time between Peacock and Babbage was fundamental. While Babbage sought to extend its formalism to all of pure mathematics—to its algebraic couching in particular—Peacock sought to ground it in an essentially nonformalist, conceptually rich and meaningful algebra.

BABBAGE'S "PHILOSOPHY OF ANALYSIS"

On 7 May 1822 Peacock wrote to Babbage to say that he had read the essays he had sent him and would return them by coach the following day. Peacock admitted to having been "completely baffled" by the latter part of the first essay, and the greater part of the second, but went on to say of the entire batch that as

> I have before expressed my opinion concerning them; they must form when completed a work of very great interest, abounding as they do with so much of original research & with illustrations of the most interesting kind.[79]

Peacock's letter registers puzzlement, disagreement, acknowledgment of Babbage's originality, and a sense of having been through it all with him before. Dubbey identifies the essays to which the letter refers as the set of papers Babbage wrote around 1821 entitled "Essays on the Philosophy of Analysis." He doubts, however, that the version perused by Peacock coincides with that preserved at the British Museum,[80] because Peacock's remarks about the length and difficulty of the first two papers "are not borne out by an examination" of the two first items on the list attached to the batch: "On notation" and "Of the Influence of General Signs in Analytical Reasoning." These, however, are *not* the first two of the papers preserved at the British Museum. The former had not gotten beyond a couple of pages at the time, and the latter, having been submitted by Babbage to the Cambridge Philosophical Society and read there by Peacock in December 1821,[81] was a paper that Peacock would have been well acquainted with long before their exchange of 1822, so it is doubtful that Babbage would have included it among the new batch of papers he had sent for his old friend's opinion. But if we assume, contrary to Dubbey, that the essays that make up the British Museum "Philosophy of Analysis" manuscript are the ones to which Peacock's letter refers, and if we take Peacock's remark to refer not to the order of titles on Babbage's roughly sketched table of contents, but to the first two essays of the batch itself, a very different picture emerges from the one implied by Dubbey.[82]

For those familiar with the relatively little written on the mathematical writings of Babbage and Peacock, the suggestion that the two men might have disagreed from the start would seem very strange. With the exception of two papers of my own,[83] everything written about the Analytical Society and its aftermath has assumed that Peacock, Babbage, and Herschel—to whom the society owes everything that was published in association with its name—had perhaps pursued slightly different topics in their research,[84] but, with regard to fundamentals, had been of one mind, and had worked in full agreement. In his interesting biographical sketch of Herschel, Schweber suggests that perhaps "all was not sweet reasonableness within the rank of the Society and that some animosity existed between the Johnian members and the others." He cites a letter from Maule to Babbage complaining of "the great difference of opinion which distract[s] the internal constitution of the society."[85] Schweber did not pursue the point any further, but gives the distinct impression that, in his opinion, the tension and "great difference of opinions" had more to do with running the society than with matters of mathematical substance.

So close are the views of Peacock and Babbage on the nature of mathematics alleged to have been that Dubbey registers surprise that the two men could have remained friends after Peacock published his *Treatise on Algebra*, which, he claims, all but plagiarizes the "Philosophy of Analysis" Babbage had shown him eight or nine years earlier.[86] Becher (1980b) goes a step further, adopting uncritically Dubbey's assessment of the two works' affinity, and attributing the fact that they remained friends to their joint debt to Woodhouse. But this is also to miss the sheer radicalism of Babbage's position, in addition to Peacock's profound motivation for, and the novelty of, his interesting departures from it.

There are, of course, obvious similarities between Peacock's mature position and the position outlined in Babbage's "Philosophy of Analysis." A closer look at the two men's early and subsequent mathematical writings indicates, however, that their initial points of departure diverged sufficiently to render the subsequent convergence of their positions impossible. Writing in 1822, Peacock gives the impression of being baffled, but not surprised by Babbage's position of which he reports to have expressed his opinion in the past. Assuming that their differences dated back at least to the late 1810s, this might very well account for the discrepancy between Babbage's memoir and Peacock's Lacroix.

Hamilton's Meta-mathematical Taxonomy

In 1837, in direct reaction to Peacock's *Treatise*,[87] William Rowan Hamilton published his oft-quoted tripartite taxonomy of approaches to algebra. Despite a certain lack of finesse, it provides an authentic, if tendentious, mapping of the field.[88] According to Hamilton, there exist three "schools" of algebra that distinguish themselves by the type of problems they address. They represent three algebraic projects, fully capable, in Hamilton's view, of peaceful coexistence.

"The Study of Algebra," wrote Hamilton in his essay "Algebra as the Science of Pure Time," which launched what would develop later into his quaternions project, "may be pursued in three very different schools, the Practical, the Philological or the Theoretical, according as Algebra itself is accounted an Instrument, or a Language, or a Contemplation." The "practical," or applied, algebraist regards algebra primarily as an art, a technique, a system of rules prized for its practical *usefulness*. The "philological algebraist"—the formalist, as I term him here—regards it as a language valued for the conciseness of its notation and its logical *coherence*. And the "theoretical algebraist" regards algebra as a science, a system of connected propositions, grounded in a clear "contemplation" of ideas and valued primarily for their *truth*.

Hamilton does not present the three views as a foundational disagreement regarding the very nature of algebra. They are not presented as being different meta-mathematical normative frameworks, but as marking a difference of mathematical *emphasis*. Algebra, Hamilton implies, is, in truth, all these things put together: an instrument, a language, and a science. It so happens that different people are attracted to the pursuit of different kinds of algebraic problems.

> The Practical Algebraist complains of imperfection when he finds his Instrument limited in power; when a rule, which he could happily apply to many cases, can be hardly or not at all applied by him to some new case . . . The Philological Algebraist complains of imperfection, when his Language presents him with an Anomaly; when he finds an Exception disturb the simplicity of his Notation, or the symmetrical structure of his Syntax . . . The Theoretical Algebraist complains of imperfection, when the clearness of his Contemplation is obscured; . . . when he . . . cannot look beyond the signs to the things signified.[89]

Ideally one would like to see one's chosen area of mathematics perfected in all three respects—as he argues is largely the case in geometry. The problem with algebra is that, despite having made considerable practical and philological progress, its theoretical basis has been neglected. In fact, algebra is now a theoretician's nightmare plagued by "confusions of thought and errors of reasoning." His delightful, almost Berkeleyan prose is worth quoting at length.

> [I]t requires no peculiar scepticism to doubt, or even to disbelieve, the doctrine of Negatives and Imaginaries, when set forth (as it has commonly been) with principles like these: that a greater magnitude may be subtracted from a less, and that the remainder is less than nothing; that . . . [two] numbers denoting magnitudes each less than nothing, may

> be multiplied the one by the other, and that the product will be a positive number, or a number denoting a magnitude greater than nothing; and that although the square of a number, or the product obtained by multiplying that number by itself, is therefore always positive, whether the number be positive or negative, yet that numbers, called imaginary, can be found or conceived or determined, and operated on by all the rules of positive and negative numbers, as if they were subject to those rules, although they have negative squares, and must therefore be supposed to be themselves neither positive nor negative, nor yet null numbers, so that the magnitudes which they are supposed to denote can neither be greater than nothing, nor less than nothing, nor even equal to nothing.

"It must be hard to found a SCIENCE on such grounds as these," he concludes.[90]

Hamilton had every right to complain about the state of "theoretical" algebra. But he was quite wrong to imply that the problem was merely a matter of algebraists getting their priorities right. The project of theoretical, or scientific, algebra was not merely suffering from lack of attention. The very idea of treating algebraic activity as contemplative, and algebraic formulas as (therefore) admitting of truth values, was under direct attack.[91] Paradoxically, Peacock, whom Hamilton took to be the representative philologist, was, at the time, more of an ally than Hamilton could have appreciated. But I am getting ahead of myself.[92]

In the terms of Hamilton's taxonomy, Babbage's "Philosophy of Analysis" represents the work of a fully fledged, unabashed, self-conscious philologist—in Hamilton's purely syntactical sense of the term. If Hamilton and DeMorgan expressed deep discomfort during the early 1830s with the idea of algebra being, as the latter would put it in his review of Peacock's *Treatise*, "something like symbols bewitched, and running about the world in search of a meaning,"[93] Babbage's "Philosophy of Analysis," written almost fifteen years earlier, argued forcefully that failure to wholly "dissever" algebra from its applications and render it an independent, self-sufficient "language of signs" (or, as he put it elsewhere, in shameless defiance of Hamilton's later taxonomy: a "science of analysis") was the greatest obstruction to its development. "The cause that has mainly contributed to fetter the language of signs," he observed, "may be found in this circumstance; that being in itself a method of reasoning of extreme generality it was discovered through the mediation of one of its particular applications: that of number."[94] Unapplied, of itself, algebra constituted a language and only a language. In Babbage's view, philologists and theoreticians were not disputing the relative urgency, but the *legitimacy* of the problems they worked on; they were disputing the very nature of algebra. From this perspective theoretical concerns regarding pure analysis were not unimportant, they were misplaced. On this point, fundamental to Babbage's position, the difference between his and Peacock's view was glaring. Where Babbage saw in the application of "pure analysis" to number a misdirected, restrictive fetter, Peacock, we shall see, saw an indispensable lifeline.

As Peacock's brief letter indicates, the "Philosophy of Analysis" represented a significant development of Babbage's earlier work, but not an upheaval. One major development his work does exhibit during this period is a move from the calculus of functions he was working on to an ambitious attempt to ground all of mathematical analysis on the principle of formal identity.[95] In terms of Hamilton's taxonomy, Babbage was determined to banish all theoretical considerations from pure analysis to its applications, thereby rendering it a matter of pure philology. Pure mathematical analysis, considered of itself, was not at all a science governed by a truth relation, but a purely syntactical system governed by a grammar. "The object which I propose to attempt," he wrote in one of the draft essays he sent Peacock, "is to separate entirely Analysis as the language of signs from all its various applications."[96] Babbage envisaged a comprehensive "science of analysis" presupposing nothing more than a series of well-defined operations subject to a formal principle of identity whenever applied. These early, unpublished works of Babbage clearly present a view of algebra that is widely thought to have only seriously begun its career in England twenty-five or so years later in the work of D. F. Gregory, the later DeMorgan, and Boole.[97] At most Babbage is credited for the role he and Herschel played in the introduction of continental mathematics into England, and for his contributions to the calculus of functions. The following remark by Elaine Koppelman is typical:

> But while Babbage and Herschel did not suggest any formal philosophy of algebra, they did in their work on the calculus of functions and in their use of the method of the separation of symbols, introduce concepts which were to be influential in the extension of the concept of algebra.[98]

"A Language of Signs"

Babbage never published such a philosophy, but he certainly formulated one. He worked hard at the "Philosophy of Analysis" for several years,[99] eventually losing interest and abandoning the project by about 1822. By then he had shown it to Peacock, who, I believe, in belated yet direct response eventually wrote the *Treatise*.

"In divesting analysis of all relation to number," he recalls in his unpublished early history of the calculus of functions (MS 13), ". . . I should consider every equation as an abridged representation of the operations which if actually performed at length would cause all the terms mutually to destroy each other."[100]

The point was made in greater detail in the "Philosophy of Analysis." Viewed thus, analysis

> ultimately resolves itself into propositions which are purely identical or at least that the signification of every equation amounts to nothing more than that when all the operations which are indicated on each side are actually executed every letter which occurs on the one side will be found occurring under precisely similar circumstances on the other and in case any letters stand as representatives for others these latter must be substituted for them before the identity becomes apparent.[101]

In Babbage's opinion, algebraic equations do not stand, of and in themselves, for "truths of contemplation." They do not *reduce* (though, of course, may be applied) to statements deemed to be true of some realm of contemplated objects or concepts. They represent formal identities regulated by operations performed upon featureless object-variables. A well-formed analytical proposition is an equation in which the two sides can be reduced to identical "letters" by means of a series of operations that Babbage apparently believed exclusively constitute the "language of signs." It is quite clear that Babbage's redefinition of analysis was undertaken in conscious opposition to what Hamilton would later describe as "theoretical" algebra and William Whewell as "pure" (as opposed to "inductive") science.[102] But it also differed, equally consciously, from other trends current among the young "analyticals" at the time. Babbage's "Philosophy of Analysis" is not concerned with developing a calculus of operations as such, nor does it seem in any way concerned with the separation of symbols (those denoting quantity from those signifying operations)—the two transitional stages credited by historians of mathematics to Babbage's generation, before the birth of a fully fledged abstract algebra in the hands of DeMorgan and Boole. Koppelman is right in noting that, like those of Herschel and to a lesser extent Peacock, "The writings of Babbage . . . testify to [his] high regard for the importance of the calculus of operations."[103] But his "Philosophy of Analysis" was geared to more than further exploiting the formal analogies or disanalogies between, say, repeated operations and the law of exponents. It insisted on the study of pure analysis in its purity, devoid of meaning, prior to application and prior to interpretation—conceiving of it, as noted previously, as what we would today term a purely syntactical system. It was not an attempt to establish *a* calculus of operations, but an attempt to distill and articulate the formal, operational system that, in his view, grounds and informs *all forms* of mathematical analysis. More than twenty years later, in an oft-quoted passage from a series of papers devoted to the foundations of algebra, DeMorgan, who is widely considered a major contributor to the development of abstract algebra, deemed such a view overly formalistic and deprived of much of what mathematics is really about. A person

> who makes the transformations of algebra by the defined laws of operation only is comparable to one who puts a dissected map together by the backs of the pieces alone; whereas the person who looks at the front, and uses his knowledge of geography to help more resembles the investigator and mathematician.[104]

For Babbage the point of foundational studies in mathematics was not merely to emulate working mathematicians, but rather to deconstruct and break down the mathematical thought process to its logically ordered elements. "It is my intention in the following essays," he wrote in the cover letter to the Cambridge Philosophical Society,

> to attempt an examination of some of those modes by which mathematical discoveries have been made to point out some of those evanescent links which but rarely appear in the writings of the discoverer but which passing perhaps imperceptibly in his mind, have acted as his unerring although his unknown guides.

The "Philosophy of Analysis" is not a study of the nature of mathematical entities. Contrary to Hamilton's concerns during the mid-1830s, Babbage's "Philosophy of Analysis" shows no direct interest in ontology. It aspires to analyze, disclose and render explicit the thought processes by which mathematical discovery is achieved.

> All that I have proposed is by an attentive examination of the writing of those who have contributed most to the advancement of mathematical science and by continued attention to the operations of my own mind to

> state in words some of those principles which appear to me to exercise a very material influence in directing the intellect in its transition from the known to the unknown.

To this end Babbage sought to differentiate sharply between the way mathematical reasoning operates in general and how it operates with respect to specific realms of objects. One of Babbage's most interesting conclusions in this respect is his observation that the relationship between mathematical analysis generally construed and its various applications is not necessarily a matter of simple interpretation. By ridding the symbols (or, as Babbage has it, the "letters") of pure analysis of all inherent reference, the domain of analysis, no longer constrained by the particularities of subject matter, is naturally expanded. But, in an important sense, he points out, the process of abstraction also acts to "contract the boundaries of the science."[105] In generalizing up from mathematical application to pure mathematical analysis not all is preserved, for in doing so the system is significantly reified. First, Babbage notes, some symbols must be restricted at least some of the time. There is the obvious need to limit some indexicals to whole positive numbers. Thus, for example:[106]

$$n=\frac{1-a^n}{1-a}+\frac{(1-a^n)(1-a^{n-1})}{1-a^2}+\frac{(1-a^n)(1-a^{n-1})(1-a^{n-2})}{1-a^3}+\ldots$$

> is only true in the case of n being a whole number and it is necessary to point out the distinction between the letter n and the letter a: the latter is perfectly unrestricted . . . all that is asserted of it is merely that when certain operations are executed, a will be destroyed by a and each of its powers will be destroyed by an equal one. With respect to n the case is different, it is essentially a whole number. The formula was probably discovered by the trial of a few of the simpler cases and the character n is merely used to signify that any number found in the natural series would give a true result. In all cases *where the meaning of a letter is confined solely to signify repetition it is to be considered as a mere number* and the origin of any formula in which such [letters] occur will easily suggest means of distinguishing them.

Similarly:

$$\frac{2\sin 2\theta}{(2\sin 2^{n+1}\theta)^{1/2^n}}=(\tan 2\theta)^{1/2}\cdot(\tan 2^2\theta)^{1/4}\ldots(\tan 2^n\theta)^{1/2^n}$$

which is only true when n is a whole positive number the identity can only be made apparent after a considerable number of steps and n must always be regarded as a mere number.[107]

But it is Babbage's next step that is the radical one, which, as we shall see further down, squarely contradicts the foundational principle on which Peacock's algebra would be grounded. "This view of the subject"—namely, to insist on the principle of strict identity—

> leads us *to exclude entirely from pure analysis all series consisting of an infinite number of terms; such expressions properly belonging to its application to number*: this does not however exclude series containing an indefinite number and in which the function containing the remainder after n terms is accounted for; thus
>
> $$\frac{1}{1-x} = 1 + x + x^2 + x^3 + \cdots$$
>
> is not legitimately an analytical expression in the sense we have assigned to the term because the expression cannot be rendered identical; but—
>
> $$\frac{1-x^{n+1}}{1-x} = 1 + x + x^2 + x^3 + \cdots + x^n$$
>
> which is a series consisting of an indefinite number n of terms may be regarded as within the province of pure signs; its identity is easily established and is perfectly independent of the nature of x: when such an expression occurs in any enquiry if a numerical result is the final object (as it most frequently is) we must make use of those rules which apply to the arithmetical interpretation of algebraic signs; the actual value of x must then be taken into consideration if n is supposed infinite and the formula will *only be numerically accurate or rather will only afford an approximation when* x *is some number less than unity.*

"From these principles," he adds,

> a satisfactory answer may be given to an objection of Berkely's [*sic*] relative to the shifting of the hypothesis:[108] as an example of it we will take the following

$$\frac{1}{1+x} = 1 - x + x^2 - x^3 + \cdots \tag{1}$$

if $x = 1$ this becomes

$$\frac{1}{2} = 1 - 1 + 1 - 1 + 1 - \cdots \tag{2}$$

which cannot possibly be numerically true; for no combination of whole numbers by mere addition and subtraction can equal a fraction (1) is therefore false when $x = 1$. Multiply it by dx and integrate then it becomes

$$\log(1+x) = \frac{x}{1} - \frac{x^2}{2} + \frac{x^3}{3} + \cdots \tag{3}$$

an expression which we conclude to be true when $x = 1$ and which from other processes we know to be accurate. Now the conclusion that (3) is true in a case in which (1) from which it is derived is false Berkely [*sic*] has justly observed is quite illogical.

His objection is completely removed by keeping the analytical process distinct from its application to arithmetic thus setting out with the expression

$$\frac{1-(-x)^{n+1}}{1+x} = 1 - x = x^2 - \cdots x^n \tag{4}$$

which is easily reduced to identity multiplying by dx and integrating

$$\log(1+x) + (-1)^n \int \frac{x^{n+1}dx}{1+x} = \frac{x}{1} - \frac{x^2}{2} + \frac{x^3}{3} - \cdots \frac{x^n}{n} \text{ or} \tag{5}$$

$$\log(1+x) + \frac{(-1)^n x^{n+1}}{(n+2)(1+x)} + \frac{(-1)^n}{n+2} \int \frac{x^{n+2}dx}{(1+x)^2} = \frac{x}{1} - \frac{x^2}{2} + \frac{x^3}{3} - \cdots \frac{x^n}{n} \tag{6}$$

Every stage in this proof is reducible to identity and is quite independent of the magnitude of n if it is required to apply it to numbers in order to find the logarithm of 2 it is to be observed that of the three terms on the left side of the equation (6) the two latter are divided by $n + 2$ on which account they become less and less as n is a larger number and when n is considered as infinite may be neglected and consequently we have as *an arithmetical form*

$$\log 2 = \frac{1}{1} - \frac{1}{2} + \frac{1}{3} - \cdots$$

In this way of stating the process there is no shifting of the hypothesis and we also see the reason why in the former process that circumstance did not introduce any fallacy for the part of the analytical formula which was omitted in the first stage although at that step of finite magnitude became by means of the integration infinitely small for the real value of *x*.[109]

FOR, AND YET AGAINST, LAGRANGE

Berkeley's "objection" was originally voiced in his infamous critique of the calculus of 1734. His attack, especially on limit-based procedures, was devastating. Increments are real quantities unless they actually vanish. Before they vanish, they are quantities that cannot be ignored in mathematics, however small. If and when they are made to vanish, they are exactly zero, and can no longer be divided by, and so forth. Berkeley saw only two alternatives: to lower standards, and view the calculus as operating with notions of rigor, truth, and exactitude less stringent than in other branches of mathematics; or to explain the obvious success of the calculus by a systematic "compensation of errors" within an otherwise rigorous mathematical enterprise. Berkeley opted for the latter. Lagrange's power series approach, so favored by Babbage and his fellow analyticals, proposed a third way, which offered to dispose of the use of limits entirely. But by choosing to meet Berkeley's "objection" by ridding pure analysis not merely of inconsistencies, but of the very notion of infinity, Babbage, it would seem, had thrown the baby out with the bath water. The Taylor series expansion of functions generates a converging, yet infinite number of terms. Lagrange's version of the calculus should, therefore, also be banished from pure analysis, not for shifting hypotheses in mid-proof, but for manipulating equations that failed to reduce to strict identities![110] Of course, Babbage could not settle for any other known version of the calculus. Any appeal to "infinitesimals or limiting ratios or exhaustions," he claimed,[111] necessarily introduced analysis into "the obscure regions of infinity" to a far greater extent than Lagrange's methods. But there is no way Babbage could have maintained, as his text implies, that Lagrangian calculus avoids the problem entirely. The way things stood, it was impossible for him to include the calculus in his scheme for pure analysis in any form. Or so it would seem.

I believe Babbage was well aware of all this. The problem was simply too obvious for him to have missed. And in any event, he solved it—quite dramatically, in fact, though without in any way insinuating that he might have found Lagrange's approach more than mildly deficient. He did so by founding the calculus anew, in a way radically different from any other, including Lagrange's. Irascible, pigheaded, direct, and rudely outspoken as he could be, Babbage, it seems, was incapable of openly criticizing Lagrange. (This makes him the only one of his generation to attempt to outdo Lagrange at his own game.)[112] He registers certain doubts indirectly, and in the most inoffensive terms, but not before paying eulogistic homage to the patron saint of British analysis, whose "masterly view" of the differential calculus "while it rested the calculus on a foundation not to be contested has at the same time furnished an example which will have no inconsiderable weight on guiding other reforms."[113] Of the great merits of Lagrange's "attempt to place [the calculus] on a better foundation" Babbage promises to speak in another part of the essay. He finally gets around to doing so, twelve tightly packed pages later, after explaining in great detail how, as in all branches of analysis, the true principles on which the calculus rests have been "obscured or infringed by a connection with any of its applications whose evidence is of a lower even of a different nature," and urging that the remedy be found in "entirely dissevering it from infinitesimals." Babbage then announces that he has a suggestion of his own, one that

> was unknown to me when I began this essay and arose entirely from pursuing the reasoning which I have explained when examining the meaning of exponents both of quantities and of functions as indicated by the two equations
>
> $x^{a+b} = x^a \cdot x^b$ and $f^{n+m}(x) = f^n f^m(x)$

And it is at this point that Babbage introduces a mildly critical comment regarding Lagrange.

> An objection to the view of Lagrange had often been proposed namely that it required that the possibility of converting all expressions into series should be granted a postulate referring to a fact which although undoubtedly true could hardly be established without considerable discussion. Besides this the very method of converting functions into series may be regarded as one of the first applications of the differential calculus and although it can be accomplished without passing through that step to which reference has been frequently made yet it still leaves the

> system open to the objection of establishing the principles by means of one of the applications.[114]

Babbage's rhetoric is amazing. On the one hand, he clearly implies that in his opinion both objections are unfounded, rendering Lagrange's move impeccable. And yet he cites them as reasons for seeking an alternative approach! And all of this, while concealing his real reason for rejecting it, namely, his categorical ban on infinite series in pure analysis!

> These reasons induced me to look amongst the properties of differentials for some one from which the others might be deduced and which should be free from the objections that have been stated. The simplest ought of course to be selected if any such could be found suitable to the purpose.[115]

His suggestion is simple and without precedent: to define "the prefixing to a function the letter *d*" as an *operation* such that:

$$d(xy) = xdy + ydx$$

and

$$d(x+y) = dx + dy$$

No limits, no infinitesimals, no exhaustions, no infinite series. Babbage's *dx* is none of these. Nor is it a "derived function" as in Lagrange's *Thories des fonctions analytiques*. It is simply an operation from which he proceeds to derive the "usual expressions of the differentials of functions" as follows:

If $y = x$

$$d(xy) = d(x^2) = xdx + xdx = 2xdx$$

If $y = x^2$

$$d(xy) = d(x^3)[= d(x^2x)] = 2x^2dx + x^2dx = 3x^2dx$$

and generally when *n* is a whole number

$$d(x^n) = nx^{n-1}dx$$

Next let it be required to find the differential of x^{-n}.

Let $z = x^{-n}$ then $zx^n = 1$ and $d(zx^n)[= d(1)] = x^n dz + nz^{n-1}dx = 0$

$$dz = -\frac{nx^{n-1}zdx}{x^n} = \frac{nx^{n-1}x^{-n}dx}{x^n} = -nx^{-n-1}dx$$

Let $z = x^a$ where a is any number[,] then it may be put into the form $z = x^{m/n}$ where m and n are whole numbers [*sic*!].

hence $z^n = x^m$ and by (a)

$$nz^{n-1}dz = mx^{m-1}dx$$

$$dz = \frac{m}{n} \cdot \frac{x^{m-1}dx}{z^{n-1}} = \frac{m}{n} \cdot \frac{x^{m-1}dx}{x^{(m/n)(n-1)}}$$

or

$$dz = \frac{m}{n} x^{(m/n)-1}dx$$

In general, Babbage concludes, all three cases are contained in the expression:

$$d(x^a) = ax^{a-1}dx$$

He goes on to note that the differential of x/y is easily deduced from the consideration of its equivalent: xy^{-1}, and at this point the essay is discontinued.

Such was Babbage's construal of pure analysis: a system consisting of well-formed operations, defined in relation to "letters" that denote mathematical objects generalized in advance to the level of absolute, featureless abstraction. Well-formed propositions within the system are represented by equations constituting statements of strict identity. Pure analysis according to Babbage constitutes an autonomous realm of mathematical activity knowingly "dissevered" from all manner of application. It is, in a sense, a meta-mathematical system, logically prior to, and hence wholly independent of, actual and potential subject matter. So pure was Babbage's pure analysis that in one important respect it went a significant step further than what would later be dubbed abstract algebra. Babbage refused to allow into the realm of pure analysis anything, as well-defined and abstract as it might be, that failed fully and rigorously to comply with the principle of identity.

Even though he wrote the "Philosophy of Analysis" several years before Cauchy published his famous epsilon-delta formulations (and failed to react to them after they were published), one may safely assume that he would not have tolerated the inclusion of inequalities parading as identities, even if they could be shown to be appropriately contained within rigorous Cauchian limits. But if the mathematics of convergent and divergent series is deemed "altogether foreign" to pure analysis, and "assigned" exclusively to "the nature of the subject on which that species of reasoning is employed," it can no longer be strictly considered an *application* of analysis in the strict sense of the term.

Babbage's pure analysis is not abstract algebra, but *rarefied* algebra. "Ordinary" algebra cannot be considered a model of the higher system that Babbage envisaged, exactly because parts of it are obtained independently. The application of pure analysis to specific realms of mathematical objects entails more than a narrowing of the scope of its expressions. It also involves concessions to rigor. In this, too, as we shall see, Babbage's construal of pure analysis differed significantly from the system Peacock would eventually propose. More important, it was the system *against which* Peacock found himself having to work as he gradually awoke to the challenge of the analytical revival of Cambridge mathematics to which he was wholeheartedly committed. The idea that their accounts of algebra converged, or that Peacock in any way plagiarized Babbage, is groundless.

CHAPTER SIX

Creatively Undecided: The Making of Peacock's Two Algebras

INTRODUCTION

The Analytical Society's well-known reformative undertaking, described in the first part of the previous chapter, was the context within which Peacock's meta-mathematical commitments acquired their initial form. They premised a full, yet contained and strictly limited endorsement of Lagrange's formalist approach to the calculus, devoid indeed of any surplus conceptual baggage in the form of limits, fluxions, or differentials, yet grounded in, and meaningful of, an essentially *non*formalist account of algebra itself. Babbage's radical, yet less-known extension and development of the society's program, portrayed in the previous chapter's latter part, served, I believe, as the crucial ambivalating factor responsible for the keen process of meta-mathematical rethinking and reformulating undertaken by Peacock, which culminated in 1830 in the detailed hybrid compromise he presented in the *Treatise*.

The present chapter consists of three main sections. The first reconstructs Peacock's point of departure as evident from his notes to Lacroix and early correspondence. The second is devoted to his attempts during the 1820s at what we might term an "emergence theory" of algebra, a depiction of modern algebra as the culmination, or end point, of a "natural history of arithmetic," to use Kevin Lambert's useful phrase.[1] As noted in passing above, this phase in Peacock's development places him firmly within the fold of what Snyder calls the Philosophical Breakfast Club, and creates an intriguing affinity, of which I had not formerly been aware, between his meta-scientific musings and those of Herschel and Whewell.

The third and main section of the chapter offers a close reading of the essentially *ahistorical* account of arithmetical and symbolical algebra as two normatively distinct, yet somehow connected realms of algebraic practice and meaning, which Peacock set forth in great detail in the *Treatise* of 1830 and his substantial BAAS Report of 1834.

I have had little so far to say about Peacock. By the time the society was founded, he was evidently someone to reckon with—one of Bromhead's initial invitees and founding member of the society, just a year or so away from being named second Wrangler and second on the Smith's Prize examination, following close on the heels of Herschel's firsts. But unlike Herschel and Babbage, Peacock seems at the time not to have been at all motivated to engage in original mathematical research. I am unaware of any mention of him presenting work at any of the society's meetings. He certainly did not contribute to the *Memoirs*, and although he was in close contact with both Herschel and Babbage, he took no part in, and seems not even to have been privy to, their intense exchanges concerning functional equations. Where lesser figures in retrospect, like Bromhead, Slegg, and Maule, tried their hand at mathematical research and communicated their findings to their friends, Peacock failed to engage.

Enros is probably right to suggest that although he excelled in his studies, at the time the society was formed, Peacock's knowledge of mathematics was no match for that of Herschel and Babbage.[2] But I doubt he lagged behind for long. His notes to Lacroix, the papers he later presented to the Cambridge Philosophical Society, and certainly the eventual *Treatise* and BAAS Report are all extraordinarily well informed, and exhibit undisputed mathematical competence and knowhow. Peacock's clear preference for meta-mathematics (like Whewell's for meta-science for that matter) cannot be easily attributed to lack of talent or capacity for first-order research. He was as zealous and dedicated a supporter of "the true faith"[3] as any of his peers, but he chose to write *on*, rather than *in* mathematics, and corresponded with his friends about it. For one thing, as the first of the society members to seriously occupy a college position—he was elected Fellow of Trinity College in 1814, appointed assistant tutor and lecturer the year after, and moderator in 1817—Peacock's initial interests focused, as would Whewell's, on the more didactical questions of curricular reform.[4] His first writings to appear in print belonged squarely within this category: (his share of) the translation of Lacroix published in 1816, and his *Collection of Examples of the Applications of the Differential and Integral Calculus*, published in 1820. Yet, again like Whewell, Peacock did not limit himself to the didactical domain proper for very long.[5] In a more general sense, however, his initial focus did remain stable throughout his career, marking yet another and more obvious difference between his approach and that of Babbage's "Philosophy of Analysis." Babbage's essays were not written with a view to making sense of algebra as he found it. They do not represent an interpretive or explanatory undertaking, at least not primarily so. They purport to articulate the foundations, formulate the objectives, and lay down the foundational principles of a new, austere, and highly original conception of mathematical analysis. They represent the work of a working mathematician—not *studies* of mathematics, but *works* of mathematics. In this respect, Peacock's writings on algebra, in both its historical and systematic phases, if as highly original, were very different.

Peacock's works set out to articulate algebra as he found it and to determine what it could, therefore, hope to achieve. His works aspired to radically *reinterpret* an existing area of mathematics; Babbage, by contrast, endeavored to construct one *de novo*. The innovations contained in Peacock's work were *meta*-mathematical. His were new expositions offering new understandings of what algebra *is* and how it is best accounted for; the novelty of Babbage's essays lay in their being a first, incomplete attempt at stating what algebra *should be*. Both projects conveyed and applied philosophies of algebra. Babbage's did so with a view to putting right something allegedly wrong *in* mathematics, Peacock's did so with a view to putting right something allegedly wrong in our *understanding of* mathematics. I stress this point because, knowing that Peacock had been privy to Babbage's essays, his decision to adopt an interpretive rather than a reformative approach suggests, as did his letter to Babbage of a few years earlier, not only a knowing disagreement with his friend's line of thinking, but an attempt to somehow incorporate whatever he found right and challenging in Babbage's proposals, within the framework of his existing commitments.

All this strongly suggests that the *Treatise* can no longer be read as an unclear and hesitant step in the direction of a formalist approach to algebra that its author wanted, but was incapable of fully seeing through. Peacock cannot be said to have been timidly groping toward a position he lacked the ability to properly formulate, for the simple reason that, as it now turns out, that very position was staring him in the face. To put it differently, in view of Babbage's "Philosophy of Analysis" and Peacock's evident knowledge of its content, the *Treatise* can no longer be set against a supposedly linear development in mathematics that allegedly culminated twenty years or so later.[6] What would appear, judging from the published record alone, as an interesting move away from one conception of algebra *en route* to another is better described as an uneasy, unstable, and short-lived attempt to jointly accommodate two mutually incompatible meta-mathematical frameworks.

ANNOTATING LACROIX

To appreciate this point, we must first briefly return to Peacock's first published work, namely, his, Herschel's, and Babbage's edition of Lacroix. Peacock's twelve notes (Lacroix 1816, 581–681) were self-consciously penned with the single aim of "translating" Lacroix's text from the language of limits to that of Lagrangian power series. As the advertisement to the book has it:

> The first twelve of the Notes were written by Mr. Peacock, and were principally designed to enable the Student to make use of the principle of Lagrange, adopting those statements and examples of our author, which do not involve the theory of limits.

"The others," it adds, "were written by Mr. Herschel."[7] Peacock's notes are detailed and demanding, but could not be otherwise if the job was to be done properly. If the coefficients of the Taylor series are to define derivatives rather than the other way around, then they need to be obtained independently of the calculus—a procedure long and arduous. It was an impressive undertaking, which must have cost its author considerable time and effort. It was also retrospectively deemed superfluous by Babbage's autobiography, as we have seen. We can now begin to understand why. Close to the time Peacock was composing his notes, Babbage was becoming more and more critical of Lagrange's Taylor series definition of the derivative—not because the Taylor series was originally yielded by means of the calculus, but because it inherently involved the use of infinite series. According to Babbage's emerging philosophy of analysis, Lagrange's version of the calculus, formalist as it appeared to be, was to be ultimately counted, therefore, not as a branch of pure analysis, but among its lower-level, applications. For Babbage, rendering a limit-based calculus Lagrangian was no longer equivalent to rendering it wholly analytical.

This is certainly true of the position Babbage had articulated by 1820 or so, and although he wrote no more on the foundations of mathematics proper, it is a position to which, we can safely assume, he remained firmly committed until the end of his life. I say so for two reasons. First, if my reading of his account of the history of the calculus of functions, on which he continued working long after publishing the *Passages*, is correct, then he is the only one of his illustrious Cambridge friends, who at the age of seventy-eight still regarded work he had essentially completed before 1820 as his main claim to posterity.[8] Second, unlike Lagrange's version of the calculus, Babbage's calculus of functions does rank, I believe, as pure analysis even according to his stringent standards. The account MS 13 offers of his pre-Cambridge and Cambridge days, I have argued elsewhere,[9] is to a large extent deceptive. Paradoxically, because of the extended attempt it makes to rewrite history, it is extraordinarily revealing of Babbage's thinking and self-image late in life. What it certainly makes clear is that, unlike Peacock and Whewell, Babbage remained throughout his life unfailingly faithful to the meta-mathematical normative framework of his youth.

From Babbage's emerging perspective, Peacock's notes of 1816 did little more than translate one lower-level application of pure analysis into another. Peacock believed his notes enabled Cambridge men to study and appreciate the calculus for what it really was. For Babbage, at least in retrospect, the aim of the textbook seems to have been to teach the calculus by means of a rigorously algebraic, functional application, and to that end, its application to limits was as good as any.

Around the time Peacock was penning the notes to Lacroix, Herschel, too, seems to have been less insistent regarding Lagrange's approach. In a section of his appendix to Lacroix, entitled "On the Connection between the Differential Calculus and That of Differences," he introduced the former liberally as "considered in the light in which Leibniz presented it, or as depending on the theory of limits" without a word of reservation.[10] I am not quite sure what to make of Herschel's attitude at that point in time. He certainly seems to have changed his tune by the time he wrote his masterly entry "Mathematics" for the *Edinburgh Encyclopedia* in 1832.[11] Peacock's position, by contrast, was both firm and clear. He stated his preference for Lagrange's version of the calculus in what Hamilton would surely have recognized as strong theoretical terms. He criticizes each of the three "methods" of the calculus separately. Of D'Alembert's "method of limits" he writes:

> Our notion, indeed, of a ratio, whose terms are evanescent, is necessarily obscure, however rigorously its existence and magnitude may be demonstrated; and its introduction into all our reasonings in the establishment of this Calculus, is calculated to throw a mystery over all its operations.[12]

The "method of Infinitesimals" is even worse, being

> liable to the same objections as the system of D'Alembert, and to many others from which that system is free; for its first principles hardly admit

> of demonstrations, and by considering the successive differentials as infinitesimals of successive orders, we are unable to form any notion whatever of their connection with each other, and with the function from which they are derived.[13]

With the method of fluxions Peacock had the least patience, deeming it "unquestionably inferior to all other methods."

> The consideration of Motion, which is essential to the method of fluxions, is foreign to the spirit of pure Analysis; and the analogy by which the name and properties of a fluxion are transferred to a modification of the difference of a function, is strained and unnatural. The different orders of fluxions also are involved in considerable obscurity, and we are utterly unable to comprehend the connection which they respectively bear to their primitive function.[14]

The problem for Peacock was not that of confusing the true foundations of the calculus with one of its applications. Nor did it stem from what Hamilton would call philological concerns. For Peacock in 1816, the problem for which Lagrange's version of the calculus was presented as a solution was one of *foundational clarity*. To ground the calculus on limits, infinitesimals, or fluxions, he insisted, was to render it conceptually obscure. He saw the great merit of Lagrange's alternative in the way it allowed the calculus to be "established upon principles which are entirely independent of infinitesimals or limits,"[15] let alone fluxions. But unlike Babbage, Peacock does not seem to have been overly motivated by the formalistic aspects of Lagrange's approach per se, or much interested in formalism as such. His notes sought less to formalize the calculus more efficiently than to better clarify its foundation with a view to enabling students to form a clearer notion of derived functions and of "the connection which they respectively bear to their primitive function."

Peacock, however, was not content to stop at that. After purging the calculus of the "aukwardness" (*sic*) of Newton's cumbersome notation[16] and establishing it anew upon the allegedly solid, algebraic basis furnished by Lagrange's definitions, he envisaged a second stage to the reform process in which, after completing their version of Lacroix, they would try their hand at a similar textbook on algebra. While still proofreading the notes, he tried to interest Herschel in his new project. "An Algebra is much wanted & so is a treatise on the application of Algebra to Geometry: the second is almost partial to a system of radical reform," he wrote in November 1816.[17] Such an Algebra, he added three weeks later, "must supersede Waring[18] & must include *a theory of numbers* & indeterminate equations: it ought to enter much more into detail than the cur[rent] English books on *the first principles*: something in short like Lacroix's, but containing more matter."[19] Thanks to Lagrange, Peacock was able to present the calculus to the English reader in all its richness and as a branch of algebra. It was now time, he felt, to get down to basics and to deal with algebra directly. Herschel, who by then was deeply immersed in functional equations and the calculus of differences, did not respond. Babbage, as far as I can tell, was not approached by Peacock, but, as we have seen, seems to have had similar plans—though not for a textbook, and certainly not for the type of algebraic foundation Peacock had in mind.

Yet, in one important respect, the two friends' initial programs did coincide. They both sought to distinguish sharply between the calculus and its applications. Neither of them denied the great value of designating limits and curvatures by means of derivatives, but both strongly objected to *defining* derivatives by means of such *applicanda*. Of itself, unapplied, the calculus, they agreed, pertained to a more basic level of mathematical "existence." Peacock located it as part of algebra; Babbage, as part of what he termed pure analysis. With the completion of Lacroix, the two men applied themselves to the task of explicating that more fundamental level. And it is here that one can discern profound disagreement, at the heart of which lay the fact that what Peacock took algebra to be in 1816 was considered a lower-level application of Babbage's emerging notion of pure analysis. Babbage's "Philosophy of Analysis" left no room at all for the type of "theory of numbers" on which Peacock sought to found algebra.

This alone would account for the marked discrepancy between Babbage's recollected plan for Lacroix and the way it was executed by Peacock. For the Peacock of 1816, Lagrange's power series rendition of the calculus provided the crucial link in the grand scheme of algebraic reduction. That is why translating Lacroix's limit-based text into the language of Taylor series coefficients was so important. For the Peacock of 1816 the original Lacroix did not speak for itself, and could not be allowed to go out into the English-speaking world unannotated. The Babbage of the "Philosophy of Analysis" could have hardly cared less! Having banished Lagrange's calculus from pure analysis, and relegating it, along with limits, to the level of application (his lip service to Lagrange notwithstanding), it no longer really mattered which application one chose in order to introduce the calculus to beginners. As noted above, even if he had thought differently in 1811 or 1812, it is easy to see how a more mature Babbage might have redescribed his initial choice of university textbook. That Peacock evidently took a very different view of Lacroix at the time is also understandable. For him, Lagrange had done for the calculus what Descartes had done for geometry: He had rendered it a branch of algebra. The next step in the process of "reintroducing the exotic" into England was only natural: to publish a work on algebra similar in standard to (the appropriately annotated) calculus of Lacroix, a project that, contrary to Babbage, Peacock believed at the time could be built upon a theory of number.

The Lockean Context

Such an evaluation of Peacock's position around 1816 not only makes good sense of the little he left us from that period, but it also fares well with what we now know, thanks to Durand-Richard's and Kevin Lambert's work on the algebraic and philosophical roots of the curious history of arithmetic project Peacock embarked on in earnest a few years later. It also fares well with Joan Richards's work on the broader educational and theological setting of early nineteenth-century British mathematics. Over and above their different notational conventions and differences in mathematical subject matter, the works of French and English mathematicians, Richards has shown, employed at the time fundamentally different modes of mathematical argumentation, geared to fundamentally different conceptions of mathematical truth.[20] While French mathematicians generally sought to ground mathematical verity upon public and formal standards of mathematical rigor, the type of mathematics understood, taught, and practiced in England pertained traditionally to a far more introspective, personalist, participatory notion of mathematical certitude. Where the French texts strove to establish mathematical certainty by applying rigorous proof procedures to exact definitions, English adaptations of the very same material sought to ground mathematical truth, much as Hamilton would urge, on a meticulous clarification of ideas. Furthermore, where the French aligned their mathematics with the public, interpersonal, demonstrative methods of the natural sciences, the British, argues Richards, appear to have aligned theirs with the more contemplative modes of (Protestant) theological reflection. Richards attributes the peculiar style of British mathematical thinking to a broadly shared and equally peculiar "evangelical" or "Pauline" notion of faith, which, she argues, was originally introduced intimately coupled to mathematics in the writings of John Locke. Locke portrayed knowledge of God and mathematical knowledge as grounded upon what Whewell would later call "progressive intuition"—personal and private processes of growing self-acquaintance.[21]

Richards's work is crucially important to understanding the wider and general cultural and intellectual background within and against which the reform process developed in England. But it is also important to the finer-grained understanding of the mindsets and subsequent development of the individuals on whom this study focuses. Snyder justly draws attention to the "Philosophical Breakfast Club" members' Baconianism. It was Bacon's methodology and grand vision of empirical science to which they indeed pledged allegiance and which they undertook to promote. But as noted in the previous chapter, Bacon's philosophy had little to offer regarding the deeper philosophical questions pertaining epistemology and truth. The works that Herschel, Babbage, Whewell, Peacock, and Jones keenly discussed and committed themselves to implementing at the Sunday breakfasts described by Snyder were those of Bacon, but the deeper-seated, normative framework of philosophical ideas that grounded that empiricist commitment was Lockean.

However, in dealing with individuals, one needs to tread more carefully, and paint with less sweeping strokes. As noted, Locke's philosophy distinguishes between ideas acquired by sensation and those acquired by reflection, by "the perception of the operations of our own mind within us."[22] The constructive portions of Bacon's methodology of science, to which all involved were then firmly committed, pertained exclusively to the former.[23] Theology and mathematics, argues Richards, traditionally pertained in England to the latter. Here, however, enough has been said for us to seriously doubt whether the four young friends also shared such commitment. With regard to mathematics, at least, a clear dividing line runs between Peacock and Whewell, on the one side, and Herschel and Babbage, on the other. Herschel's commitment to Locke's sensationalism (and to Bacon's philosophy, of course) was, and was to remain, firm, but I doubt that he or Babbage had at any point developed a comparable attachment to the kind of Lockean mathematical "intuitionism" of which Richards speaks. On the contrary, once Herschel had overcome his initial aversion or indifference to mathematics, and Babbage his eclecticism, everything they wrote and did attests to a firm and uncomplicated commitment to the most formalistic brand of what Richards identifies as the French "style" in mathematics.[24] In his "Philosophy of Analysis" Babbage, we have seen, took this position to the limit, but from that point on he seems to have been sufficiently at peace with his meta-scientific convictions not to have felt a need to address foundational issues ever again. Herschel also remained faithful to both sets of his initial commitments: to Bacon and Locke with respect to "natural philosophy," and to the view of mathematics he shared with Babbage. But because, unlike Babbage, he worked and thought and wrote intensely in and especially about both areas, Herschel, I believe, did become aware of how uneasily they sat together, but unlike Whewell or Peacock, he preferred to let sleeping dogs lie and left the tension unattended.

Whewell, who believed that for the reform process to succeed, the "new math" had to be proven more effective than the old in solving problems, especially in mechanics, devoted the first years of his career to writing two major mechanics textbooks for the university.[25] And it is here, in the context of contemplating the didactics of teaching mathematical physics, that, I have argued,[26] he first awoke to the difficulties of squaring a Baconian-Lockean, realist conception of physics with the Lagrangian view of its mathematical couching favored by the society. But this was only the first step in a long journey of meta-scientific self-reckoning. By the mid-1830s, Whewell, by far the keenest and most philosophically acute of the four friends, would find himself forced to partly rethink both components of his initial set of commitments, in favor of a semi-Kantian alternative he never succeeded in fully accommodating. However, in one major sense Whewell's impressive philosophical undertaking fell disappointingly short of both its ambition and promise. He never succeeded in matching his admirable and highly original philosophy of "inductive" science with a comparable philosophy of mathematics (although, as we shall see in chapter 7, he seriously tried), because he found it impossible to accommodate the "higher analysis in mathematics" within the essentially Lockean framework to which he was committed. Whewell never seems to have considered either of the two possibilities to which Peacock would devote the lion's share of his efforts: viewing them as successive stages in a meaningful, progressive development within mathematics, or as existing side by side in meaningful symbiosis.

Problems Accentuated

Peacock's initial commitment to Lagrange's version of the calculus was obviously firmer and deeper than Whewell's, and he would remain faithful to it for the rest of his career. Yet he was also, and would also remain, as firmly committed as Whewell was at the time to a fundamentally Lockean account of mathematical meaning and truth—which as far as mathematics was concerned, renders him the most interesting of the four friends.

As Richards's work clearly demonstrates, such a commitment ran deepest at Cambridge, where it played a far more significant pedagogical role than it did at Oxford, for instance.[27] In the light of Richards's analysis, it is both significant and understandable that Herschel and Babbage, the two most ardent early spokesmen for the opposition, would encounter serious difficulties adapting to Cambridge, and end up operating from outside the university for most of their lives.[28] It also explains why Peacock, who was initially so determined to accommodate Lagrange within a Lockean framework, took his position at the university so seriously. But by the time he became fully exposed to Babbage's "Philosophy of Analysis," Peacock realized that the task he had set himself posed a challenge exceedingly greater than that of accommodating Lagrange within a workable and well-established algebra conceived of as a "science" of number or quantity. The problem that was driven home forcefully by his exposure especially to Babbage's draft essays was: Not only did such an algebra not exist, but there was no clear way of seeing how it could!

The discrepancy between what was being done in, and with, algebra at the time and the level, especially in England, of Lockean committed attempts to articulate and explain the nature and foundations of algebra was enormous. For someone as concerned as Peacock evidently was with what Hamilton would term the "theoretical" aspects of algebra, Hamilton's complaints against the sorry state of such studies were as true in the late 1810s as when they were issued in the mid-1830s. At the time, the only conceivable theoretical option open to Peacock was to ground algebraic truth in some form of generalization of the notion of number. So far, however, all such attempts had failed dismally to provide an algebraic undergirding capable of supporting anything like Lagrange's version of "Euler's calculus." In later years, Peacock would cite the works of "Lockean mathematicians"[29] Francis Maseres and William Frend as notorious examples in this respect. Both writers had raised the problem of treating negatives and imaginaries as species of quantity, and both had proposed, incredibly, to resolve it by denying them a place in algebra altogether. In this way algebra was construed as a universal or generalized arithmetic.[30]

In the *Treatise* and especially in the BAAS Report Peacock was forced to agree with Maseres and Frend that "arithmetical algebra" is "the only form of [algebra] which was capable of strict demonstration, and which alone, therefore, was entitled to be considered as a science of strict and logical reasoning." The arguments of the two British algebraists, he concludes, "were unanswerable . . . and the system of algebra which was formed [devoid of negatives and imaginaries] . . . was perfectly logical and complete."[31] By this I take him to mean that only thus did he think that algebra could be considered a science (in Hamilton's sense of the term) susceptible to a truth relation (or, as he put it in the *Treatise*, as a system founded upon axioms that are considered true in addition to definitions that are considered well formed).[32] But he could not agree with Maseres's and Frend's outrageous conclusion to dispense, therefore, with the "great multitude of algebraic results and propositions, of unquestionable value and of unquestionable consistency with each other, which were irreconcilable with such a system, or, at all events, not deducible from it . . . and which made it necessary to consider negative and even impossible quantities as having real existence in algebra, however vain might be the attempt to interpret their meaning."[33]

If in 1816 Peacock's chirpy letter to Herschel could suggest that he was seriously toying with the idea that modern analysis, Lagrange-style, as he knew it, could be fully couched in the type of meaningful, truth-governed, Lockean algebra to which he was committed, it is clear that by 1830, Peacock had despaired of establishing a science of number rich enough to accommodate the entire system. By then he had come to realize, as he would put it, that algebra consisted of two algebras, two separate and distinct, yet interestingly related realms of algebraic meaning and practice: arithmetical algebra, more or less identical with that of Frend and Maseres; and symbolical algebra, which resembled, but, *pace* Dubbey and Becher, was by no means identical to, Babbage's conception of analysis.[34]

If Peacock indeed underwent an early, naive phase of the kind of high hopes implied by his letter to Herschel, it could not have lasted for long. In 1816 he already knew enough algebra and enough about algebra to know full well that the kind of system envisaged by Frend and Maseres was incapable of providing the foundations for the modern mathematics he endeavored to promote at Cambridge. Clearly, at that point in time, like Whewell, though contrary to Babbage and Herschel, he was deeply and firmly committed to a squarely "theoretical" conception of mathematical meaning, knowledge, and truth, late of Locke. He would still have found the conclusions of Maseres and Frend outrageous, but there is much to indicate that he had not yet despaired of seeing through a more sophisticated version of the type of truth-governed algebra they had taken for granted.

I have come across nothing that bears directly on Peacock's deliberations between 1816 and 1830. What we do have is a good idea of his options around 1820 or so, and what he ended up publishing from 1829 onward. The options he faced in 1820, we now realize, posed a cruel divide: Maseres's and Frend's "generalized arithmetic" approach, on the one hand, *versus* something akin to Babbage's pure analysis, on the other. The former proposed sacrificing whole realms of algebraic activity for the sake of preserving for mathematics a solid, meaningfully Lockean, truth-governed basis. The latter insisted on draining analysis of all meaning and content for the sake of grounding mathematics anew by means of an abstract system pertaining exclusively to pure form and process. The two frameworks were, in principle, incompatible. It is not hard to imagine Peacock's difficulties with the former, which were to be clearly stated in the passages of Peacock 1833 critical of Maseres and Frend, cited above. The problems he might have had with Babbage's radical formalism are a different matter. He never addressed such an approach publicly and therefore never voiced his misgivings. By the time his mature position was taking form during the late 1820s, Babbage, who had abandoned his "Philosophy of Analysis" unfinished, was no longer a player. In what he published thereafter, Peacock appears to have decided against openly confronting a system of thought its author had decided not to make public and of which his readers could not be aware.

Lacking Peacock's own testimony, we can only speculate. But, as indicated above, there are good reasons to assume that he embarked on his foundations of algebra project sufficiently committed to a "theoretical" framework to rule out in advance any possibility of a clean and total conversion to Babbage's thoroughly formalist alternative. At the same time, we can also assume that he was sufficiently skeptical of the "official" Cambridge, Lockean version of the former to realize that much of its rhetoric was mere lip service to a program that with regard to algebra as he knew it, was quite incapable of being realized. This is the context in which, I believe, the long process that culminated in the *Treatise* should be read—a process that comprised two distinct stages that yielded two major and extraordinarily impressive, yet very different works.

PEACOCK'S NATURAL HISTORY OF ARITHMETIC

In his letter to Babbage of 7 May 1822,[35] acknowledging having read his "Philosophy of Analysis," Peacock adds the following: "I have before expressed my fears that some difficulties would arise in publishing them in the Cambridge Transactions," and suggests that Babbage would do better to offer the work to the university press. However, he added, the Cambridge Philosophical Society would "be proud of receiving them from time to time as separate essays."[36] Peacock's reservations about submitting such a radically formalist work for publication by the newly founded (1819),[37] innovative, yet strictly establishment Philosophical Society, were well founded in the light of the strong Lockean leanings of even the university's most innovative faculty. But Peacock was equally serious about Babbage presenting individual essays at the society's meetings nonetheless. Indeed, between May 1820 and December 1825, four of Babbage's papers were presented to the Cambridge society, the first three of which were later published in its *Transactions*. Babbage only read the first of them. The latter three were read to the society by Peacock in Babbage's absence.[38] But it is clear that shortly after his exchange with Peacock in 1822, Babbage's interest in articulating the foundations of mathematics and in seeing the "Philosophy of Analysis" through the press had considerably waned. The two last papers Peacock read for him at Cambridge were technical and in no obvious way part of that project at all.

Peacock's interest in foundations, in contrast, not only intensified during those years, but seems to have taken a surprising turn that lent his commitment to Bacon and Locke a new and fascinating, two-sided dimension. I say "seems to have taken" because no text or letter of Peacock's has so far come to my attention in which he reflects on, or attempts to explain, what he was trying to achieve. I suspect, as I shall speculate below, that he had talked long and hard about the new project with Whewell—friend, fellow tutor at Trinity, and major contributor to the Cambridge Philosophical Society—who would pursue an interestingly similar trajectory in his own work a decade later. They most probably talked long and hard about the project, but unfortunately left no written record of their reflections. Still, a close examination of the impressive work Peacock produced during those years against the backdrop of the meta-mathematical predicament in which he was caught suggests a very interesting undertaking.

Between 27 February 1826 and 17 February 1828, Peacock presented eight(!) papers at meetings of the Cambridge Philosophical Society, but submitted none of them for publication in its *Transactions*. The first two (read 27 February and 13 March 1826, respectively) were on Greek arithmetical notation. The third and fourth (10 and 24 April 1826) dealt with the origin of Arabic numerals and the date of their introduction into Europe. The fifth (11 December 1826) concerned the numerals of the South American languages. And the last three (12 and 21 March 1827 and 17 February 1828) were devoted to Egyptian hieroglyphical symbols. The reason Peacock did not submit his papers for publication individually was that, not unlike the essays composing Babbage's "Philosophy of Analysis," his also were conceived as parts of a larger, comprehensive work: the extraordinarily rich and detailed entry "Arithmetic" on which he was working for the first volume of the *Encyclopedia Metropolitana*, published in 1829.[39] It was an enormous undertaking, consisting of over 150,000 words packed into over 150 double-column, tightly printed pages, that surveyed and analyzed a staggering array of methods and languages of counting, past and present, and mapped and charted, summarized and exemplified the development of calculating techniques and their symbols.[40]

In their studies of this phase of Peacock's work, Durand-Richard and especially Lambert have richly contextualized the arithmetic project, showing it convincingly to have been very much a child of its place and time.[41] They point especially to its distinctly Lockean nature and to the central role played at Cambridge by Locke's philosophy in general and theory of language in particular.[42] The picture painted by Lambert is finer grained and better focused on Peacock's specific case than Durand-Richard's more general portrayal. It nicely distinguishes between radical and less radical eighteenth- and early nineteenth-century appropriations of Locke, and their relation to the "Lockean mathematicians" referred to above—algebraists Frend and Maseres. Lambert also makes a clear and compelling case for how well Peacock's "natural history" approach to arithmetic tallied with the stated objectives of the recently founded Cambridge Philosophical Society, in which he and Whewell were so centrally involved. Taken together, their studies paint a striking picture of the empiricist and inductivist view of language, truth, and mathematics late of Bacon and Locke that grounded Peacock's normative point of departure.

But Lambert also understands that Peacock's arithmetic represented much more than an authentic product of its institutional and intellectual settings. The commitments that ground it were certainly Lockean—if, as Lambert argues convincingly, of a decidedly more historical, philological variety than those of Frend and Maseres—and as such certainly true to the Cambridge of their day. But Lambert, who takes interesting issue with my former work on Peacock, is also well aware of the personal, deliberative factors that motivated its author to take action—namely, how forcefully those very commitments were challenged in Peacock's case by those represented by Babbage's "Philosophy of Analysis," and how the ambivalating force (my term, not Lambert's) of their fundamental incongruity was what motivated Peacock to thoroughly rethink the genesis and development of algebra from (Lockean) scratch.

However, both researchers firmly insist that Peacock's "Arithmetic" entry and *Treatise* of 1830 should be viewed as two successive, mutually reinforcing parts of a single, sustained, and coherent whole—that not unlike Whewell's *History of the Inductive Sciences* (1837b) and *Philosophy of the Inductive Sciences* (1840), Peacock's impressive survey of the genesis and development of arithmetic was meant to yield arithmetical algebra as we know it, which in the *Treatise* was subsequently shown to similarly yield (by suggestion) a symbolical algebra akin to that proposed by Babbage. It is a tempting option, but I beg to disagree. A closer look at the two works in the light of an appreciation of Peacock's predicament reveals, I believe, a different and far more interesting story.

Building Bottom-Up *versus* Reasoning Top-Down

The most important point to realize is that for someone as deeply committed as Peacock to a Lockean, "theoretical" view of algebra, yet, at the same time, as aware as he was of the fundamental shortcomings of the generalized arithmetics of Frend and Maseres, his immediate objective would *not* be to justify a purely symbolical algebra of the kind Babbage proposed, but to make the case for a richer *arithmetical* algebra than that of Frend and Maseres—a science of number (as opposed to a purely formal system of rules) arguably rich enough to accommodate negatives, imaginaries, and all other extensions of quantity dealt with in modern analysis. Lambert puts it concisely: "Peacock's problem," he writes, "was not statements like $2 + 2 = 4$ but rather $2 - 4 = -2$."[43] But Lambert is quite wrong to rephrase this as claiming that, therefore: "The problem for Peacock was not justifying the universality of basic arithmetic but the generality of symbolic algebra." A committed Lockean would have no interest in justifying the claim that mathematical certitude rested on vacuous formal definitions. Indeed, the statement $2 - 4 = -2$ is *not* a statement of symbolical algebra at all, but a statement of an enriched arithmetical algebra! Peacock's problem in both his arithmetic project and the eventual *Treatise* was one and the same: *to justify an extension of traditional "Lockean algebra" capable of accommodating negatives, imaginaries, and so forth* (similar to what Hamilton sought to achieve when he set forth on his "Algebra as the Science of Pure Time" project). Peacock embarked on his arithmetic project, I submit, with a view to doing so directly, by reasoning up, diachronically from a detailed study of human numerical practices, languages, and techniques to a sufficiently rich generalized arithmetic. But the project failed, I believe.

The *Treatise*, by contrast, aspired, as I shall read it, to do so by going *in the opposite direction*. Rather than offer a Lockean, bottom-up argument from numerical languages and arithmetical practices to directly yield a form of generalized arithmetic richer than what Frend and Maseres had to offer (as I believe the arithmetic project aspired, but failed to do), the *Treatise accepted* Frend's and Maseres's conception of algebra—which it terms "arithmetical algebra"—as its point of departure. It then endeavored to abstract up and away from arithmetical algebra thus construed, in the distinctly *non*-Lockean fashion Peacock dubs "suggestion," to yield a totally formal, Babbage-style algebraic system, emptied completely of all traces of numerical or quantitative meaning—a formal system of rules the *Treatise* calls "symbolical algebra." The third step consists in applying results obtained symbolically back down again to enrich and extend arithmetical algebra's traditional number-based semantics.

As I shall argue in detail in the following section, the *Treatise*'s two-way argument yields a hybrid conception of algebra that is problematic to the point of mathematical incoherence, certainly from the kind of Lockean perspective to which its author was committed. In this I also disagree with Durand-Richard and Lambert, who both view it as a viable solution. What cannot be denied, of course, is that *Peacock* deemed his solution to be both viable and satisfactory. But even that, we shall see, did not last for long.

First, however, to the arithmetic project and to my reasons for insisting that Peacock's two great undertakings at least ended up largely unrelated.

Had Peacock had nothing better to do during the early 1820s, had he been idly casting around for an interesting project, then given what we know about the Cambridge context at the time and his own Lockean leanings, an attempt to provide the groundwork for a full-blooded inductive, Lockean account of the emergence of arithmetical algebra and its grounding in universal counting practices would have made perfect sense. But what we know about Peacock's prior involvement in the Analytical Society, his commitment to the reform of Cambridge and to Lagrange's version of the calculus, the urgent need he saw for an account of algebra capable of sustaining it, and his inability to accept Babbage's formalist alternative: all strongly militate against the idea that all he had set out to do in his history of arithmetic papers was merely to justify Frend's and Maseres's account. What could such justification prove? In what way could it help him address the problem of providing a viably "theoretical" account of algebra as the science of number and quantity capable of accommodating Lagrangian analysis? It was not as if Frend's and Maseres's account of algebra required justification. On the contrary, if Peacock had harbored any hopes with regard to their work, they were to surpass it! And to surpass it in true Lockean fashion was to go back to the drawing board and see whether a more thorough, more systematic, and more carefully analyzed survey of human counting practices, numeric vocabularies, and calculating techniques was capable of yielding a Lockean account of algebra *better* than theirs!

Given Peacock's particular circumstances and predicament, no other explanation of the enormous effort he decided to invest in arithmetic makes sense. It is possible to disagree that this was indeed the intended objective of the arithmetic papers—a point to which I shall return shortly—but one cannot deny that *if* it was, it was a project that ended in failure. Peacock did not succeed in reasoning up inductively from the counting and calculating techniques he studied to an arithmetical algebra capable of accommodating $2 - 4 = -2$.

Durand-Richard and Lambert argue—the former implicitly, the latter explicitly—that the idea of the arithmetic project was not to yield a richer science of number or quantity *in or of itself*, but to serve as the basis for the project of the *Treatise*. On such a showing, it was meant to inductively yield no more than an account of arithmetical algebra akin to the generalized arithmetic of Frend and Maseres, which would then be shown to yield "by suggestion"—a term I shall explain in the following section—a wholly formal and unconstrained calculus of algebraic operations that rested, similarly to Lagrange's calculus and Babbage's pure analysis, on formal definitions devoid of any notion of number or quantity. The various formulas of "symbolical algebra"—the name Peacock chose for algebra's "suggested" formal component—because unconstrained, are then employed to enrich arithmetical algebra by applying them back down to the realm of number to yield negatives and imaginaries and the other unruly forms of quantity arithmetical algebra was incapable of yielding of its own accord. According to Lambert, Peacock's "Arithmetic" and *Treatise* cohered to jointly offer a powerful justification of symbolical algebra, by presenting a comprehensive Lockean account of its emergence, clinched by what Lambert dubs a "philosophy of suggestion" (which, one might say, supposedly stands to his history of algebra as Whewell's philosophy of induction stands to his history of science). "The science of counting had suggested arithmetic," claims Lambert, "arithmetic had suggested arithmetical algebra, and finally arithmetical algebra had suggested symbolical algebra."[44]

But such an account suffers from at least two major flaws. The first concerns the notion of suggestion itself, of which I shall have more to say in the next section. At this point, suffice it to say two things: first, that it is used in the *Treatise* as a purely *formal* relation between equations, or as Peacock calls them, "equivalent forms." The laws of symbolical algebra are yielded by emptying those of arithmetical algebra of all numerical content. The formulas of arithmetical algebra denote quantitative equalities. When emptied of content and rendered "symbolical," quantitative equality morphs into formal equivalence or identity. In arithmetical algebra $a = b$ means that a and b are *numerically* equal, that they represent identical *quantities*. In symbolical algebra it means that the algebraic form a is interchangeable with b not because they are equal, but because each yields the other. This, however, is *not* the relationship between "the science of counting" and arithmetic, or that between arithmetic and arithmetical algebra. The step that gets Peacock from arithmetical to symbolical algebra in the *Treatise* is categorically different from those that typify the Lockean ascent from counting to generalized arithmetic, in which numerical content is not only preserved, but deepened. Ampliative, inductive Lockean ascent is not a formal process at all.

The second point about suggestion is that the term *suggestion* itself is Peacock's own coinage. Had he believed that each stage in the ascent from counting suggested the one that follows, as Lambert maintains, one would have expected him to say so. But the word *suggest* is not used at all throughout the arithmetic project in a sense even remotely similar to its use in the *Treatise*. The notion of "a science of suggestion" appears in Peacock's writing on mathematics as exclusively characterizing the way arithmetical algebra stands to symbolical algebra, namely, only in the *Treatise*.

Finally, even if one rejects Lambert's liberal use of "suggestion," the very idea of viewing Peacock's two projects as jointly composing a single, sustained Lockean effort is seriously undermined by the way Peacock actually presents them. For one thing, the arithmetic project does *not* culminate in the starting point of the *Treatise*. Conjoining the two projects simply does not yield the nice sequence of seemingly consecutive stages "that culminated in nineteenth-century Britain," as Lambert describes it.[45] The progression of the historical chapter of the arithmetic project (Peacock 1829, 372–482) culminates not in arithmetical algebra at all, but in what Peacock terms "the two great divisions of the science of arithmetic" (483): first, in "Arithmetic Part I" (483–503), a detailed exposition of the rules of calculating and standard notation of addition, subtraction, multiplication, division, and the extraction of roots; and then in "Arithmetic Part II" (504–23), their application "to the solution of such classes of questions as arise in the business of life" (483), such as calculating proportions and percentages with respect to quantities expressed in concrete units like dates, weights, and money. The next step in the progression, from arithmetic to arithmetical algebra, is not mentioned or even hinted at!

When arithmetical algebra does make its appearance in Peacock's work, namely, in the *Treatise*, it does so as a freestanding science in a manner wholly unrelated to the historical process by which it seemingly came into being, or to the set of practices on which it is based. Indeed, not only is arithmetical algebra not presented in the 1830 edition of the *Treatise* as the product of a historical process, its relationship to symbolical algebra is not presented there as an additional last step in such a progression, as

Durand-Richard and Lambert would have it. The aim of that work, we shall see, was *not* to claim that, once successfully suggested, algebra *became*, or *evolved* into, or was *subsumed under* symbolical algebra. For the *Treatise* of 1830, arithmetical algebra was not a ladder by which to reach a higher level, which, once achieved, could be discarded (as was the case with respect to all the various stages of Peacock's history of arithmetic). Peacock's symbolical algebra was not derived in order to *replace* arithmetical algebra, but in order to appropriately *enrich* it by reasoning back down to it, as explained above.

In sum, Peacock's notes to Lacroix and letters to Herschel clearly imply that during the late 1810s and early 1820s he saw as his most urgent task that of articulating an account of algebra *as a science of number* that viably transcended the draconian constraints placed on such a theory by Frend and Maseres—a task rendered all the more urgent by his exposure to Babbage's "philosophy of analysis." I seriously doubt that Peacock would have embarked on a project of as daunting a scope as his "Arithmetic" merely for the sake of supplying Frend's and Maseres's account of algebra with a grounding in human practice rather than in introspection, as Lambert implies. The natural history of arithmetic failed to yield its desired outcome. It could have easily culminated in an account of algebra as generalized arithmetic, but the extra effort would have served no purpose. Peacock certainly deemed it to be a valuable study worthy of publication in itself, yet, at the same time, to be one that, contrary to his expectations, had disappointedly proved in retrospect to be irrelevant to his main and most urgent objective. To this end he embarked on the *Treatise*, to which we now turn, in a manner that was largely unconnected and unrelated to the arithmetic project. This is why I believe the two projects not only failed to connect, but consequently so completely and conspicuously ignore one another.

PEACOCK'S TWO ALGEBRAS

Unlike his "Arithmetic," Peacock's *Treatise on Algebra* of 1830 was an extraordinarily influential work—not in the usual sense of the term, but in the sense developed in the first part of this study. It is not that people found his account of algebra convincing and adopted it. Although no mathematician of standing comes to mind who did not take it in all seriousness—yet at the same time, none comes to mind who can be easily classified as having ended up a true follower of Peacock (including, as I shall argue, D. F. Gregory and the later Hamilton, who both pledged allegiance to his position). And yet, despite convincing no one, it left none unchanged. But to understand its impact on others requires us to read it in the light of *their* commitments, deliberations, and personal and professional circumstances. Indeed the variety of reader responses to the *Treatise* is quite extraordinary. Whewell and Hamilton, whose positions on matters meta-mathematical and meta-scientific were near identical, still read the *Treatise* very differently, as we shall see. Whewell wrote a book-length study of the "pure sciences" of his own under the impression that Peacock's algebra lent itself naturally to a "theoretical" interpretation, while Hamilton considered it the very paradigm of algebraic "philology" and, in direct response, began fashioning his own "theoretical" alternative.

First we must consider the precise nature of its undertaking. The *Treatise*, I submit, is a work that aimed to seriously abate the intense and fundamental normative ambivalence that Peacock experienced as a result of failing to adequately meet the challenge of Babbage's decidedly non-Lockean approach to the foundations of mathematics from within a Lockean framework. And yet, as we shall see, it was a work that unwittingly addressed that ambivalence by forcefully preserving it, and in a sense accentuating it in ways it had never previously been articulated.

In view of the arithmetic project's failure to yield a Lockean algebra rich enough to ground modern analysis, the solution proposed in the *Treatise* would seem both elegant and simple. Since it proved impossible to present algebra in all its complexity as a fully fledged science of number or quantity (as Hamilton would also try to do, and would also ultimately fail), and given Peacock's reluctance to follow Babbage in viewing it as a thoroughly formal system, the *Treatise* proposed something akin to a Solomonic compromise: to split algebra down the middle, as it were, presenting it as comprising two related, yet antithetical components. One component was a truth-governed science of number that he called *arithmetical algebra*, extending as far as possible; the other, a purely formal, symbolical system that he dubbed *symbolical algebra*, unconstrained and autonomous, and sufficiently general to accommodate the full range of algebraic expression. It envisaged the former, as noted, as being much the same as the generalized arithmetic advocated by Maseres and Frend; the latter, at first blush at least, as being much the same as Babbage's pure analysis.

The problem, of course, was to determine the precise relationship between the two. Babbage would not deny there being a science of number or quantity, any more than he would deny the existence of geometry or mechanics. What he objected to was granting any of them a foundational role for pure mathematics. Babbage treated pure analysis as a freestanding formal system capable of being effectively *applied* to the realms number, space, motion, or force. To borrow the terminology of the day, Babbage, one could say, regarded arithmetical algebra and geometry as areas of "mixed" mathematics, as applications of pure analysis, on a par with mechanics and optics. Given his prior commitments, Peacock could not agree to viewing the theory of number or quantity merely as a byproduct, or a lower-level application of its formal counterpart. To some extent, in some way, symbolical algebra had to be shown to meaningfully *depend* upon, *pertain* to, if not actually *derive* from a theory of number. And yet this had to be done without allowing symbolical algebra to lose any of its unbounded generality.

Still, the very recognition of an inherent two-sidedness *within* algebra was, in itself, a significant concession to Babbage on Peacock's behalf. The opening sentences of his preface to the *Treatise* clearly convey his sense of mission and urgency, but most of all, the extent to which the difficulties he had faced required rethinking fundamentals.

> The work which I have now the honour of presenting to the public was written with view of conferring upon Algebra the character of a demonstrative science, by making its first principles co-extensive with the conclusions which were founded upon them: and it was in consequence of the very particular examination of those principles to which I was led in the course of this enquiry, that I felt myself compelled to depart so very widely from the form under which they have been commonly exhibited. The object which I proposed to effect is undoubtedly one of great importance, and of no small difficulty . . . and I am very sensible of the great responsibility which I incur by an attempt of this nature.[46]

In theory, Peacock had two options. He could divorce his two algebras completely and thus avoid having to relate them altogether. On such a view, arithmetical algebra would remain an autonomous, contemplative science of number, grounded exclusively on its own principles, no longer considered an application or model of a system more fundamental. And symbolical algebra, truly unconstrained, could be developed, Babbage-style, wholly "dissevered" from all lower-level, subject-orientated systems.[47] The second option open to Peacock was to retain a sufficiently watered-down, bottom-up dependency of symbolical on arithmetical algebra that would somehow permit meaningfully grounding the former in the latter, yet without imposing upon its generality.

Not surprisingly, Peacock opted for the latter. He did so, as noted, by assigning to arithmetical algebra the role of what he termed a "science of suggestion" for symbolical algebra, treating arithmetical algebra

> as the science, whose operations and the general consequences of them should serve as the guides to the assumptions which become the foundation of symbolical Algebra: thus granting or assuming in the first instance the universality of the values and of the representations of the symbols . . . assumptions to which there is nothing corresponding in arithmetic.[48]

At the outset Peacock makes it all sound innocent enough. He repeatedly stresses the fact that the two algebras

> under no view of their relation to each other, can be considered as one science, whatever be the nature of their connexion with each other; . . . there is nothing in the nature of the symbols of [symbolical] algebra which can essentially confine or limit their signification or value.[49]

Symbolical algebra, Peacock promises time and again, is to stand on its own two feet. Although its principles may be suggested by arithmetic, once formulated, they are to establish "the science of algebra" as "a science of symbols and their combinations, constructed upon its own rules, which may be applied to arithmetic and to all other sciences by interpretation" (1834, 194–95). In this crucial passage, Peacock goes to great lengths to play down the significance of the role he assigns to arithmetical algebra. The principles of symbolical algebra may be suggested by those of arithmetical algebra, but on no account can they be considered generalizations of them.

> [I]t is an abuse to the term *generalization* to apply it to designate the process of mind by which we pass from the meaning of $a - b$ when a is greater than b, to its meaning when a is less than b, or from that of the product $a \cdot b$, when a and b are abstract numbers, to its meaning when a and b are concrete numbers of the same or of a different kind. (194)[50]

The fact that, despite their initial suggestive role, the principles of arithmetical algebra are ultimately considered interpretations of those of symbolical algebra imposes a firm logical "order of succession" between the two realms of algebraic activity. Logically speaking, symbolical algebra needs to be established prior to interpretation, and hence prior to its arithmetical counterpart. "Interpretation," as he puts it, "will *follow*, and not *precede*, the operations of [symbolical] algebra" (195). But Peacock recoils immediately. Apparently uneasy with the idea of relegating the role of the idea of number for algebra to that of a mere inspiration, the very next paragraph takes away much of what the previous one had just given. If one is aware of the fundamental nature of Peacock's predicament, this wavering or dithering of his normative ambivalence is manifest almost from page to page.

It is not enough that the principles of symbolical algebra be merely inspired or suggested by those of arithmetical algebra. In the very next paragraph Peacock goes on to demand that they be so constructed as to ensure *in advance* their capacity to yield their arithmetical counterparts by interpretation. The principles and rules of combination of symbolical algebra, he insists, are to be preserved "in whatever manner they may have been suggested." Arithmetical algebra is required, therefore, not only to suggest symbolical algebra, but to be included in it as a special case. Notice also how the move from mere suggestion to *predetermined inclusion* is subtly clinched by changing the idiom to that of necessity.

> Though the science of . . . arithmetical algebra does not furnish an adequate foundation for the science of symbolical algebra, it *necessarily* suggests its principles, or rather its laws of combination; for in as much as symbolical algebra, though arbitrary in the authority of its principles, is not arbitrary in their application, being required to include arithmetical algebra as well as other sciences, it is evident that their rules must be identical with each other, as far as those sciences proceed together in common. (195)

The principles of symbolical algebra are not merely to be motivated by those of arithmetical algebra, but are required a priori strictly to embody them.[51] This is Peacock's first, and explicit, step toward denying symbolical algebra the autonomy and self-sufficiency he had so ceremoniously promised at the outset. Still, visibly dithering, he continues to insist that

> [t]he rules of symbolical combination which are thus assumed have been *suggested only* by the corresponding rules in arithmetical algebra. They cannot be said to be *founded* upon them, for they are not *deducible* from them. (197–98)

It is true that in Peacock's system the principles of symbolical algebra are not *deducible* from those of arithmetical algebra, but they are nonetheless wholly *determined* by them. His argument to the contrary, cited a moment ago, rests on the claim, dubious in itself, that the leap from contemplating the subtraction of a lesser from a larger quantity to the subtraction of any quantity from another is so great as to do violence to the term *generalization*. I think most people would disagree, as I certainly do, but that is beside the point. The problem with this line of argument, central to Peacock's case, is not that it is invalid, but that it is quite irrelevant to the problem at hand. The move, at the level of principles and rules of operation, from arithmetical to symbolical designation does *not*, at any point, require the mind to entertain the idea of subtracting larger from lesser quantities. The very idea of a symbolical algebra, founded as a "science of symbols independently of arithmetical considerations," is *not* to expand and generalize the notions of number or quantity, but to get away from them altogether! As I argued against Lambert, above, in symbolical algebra (as opposed to generalized arithmetic), prior to its application, the a and the b in $a - b$ are utterly featureless and, in and of themselves, designate nothing in particular. They do not stand for quantities lesser or greater, because they do not stand for quantities at all! They stand for nothing; they are merely symbols, or "letters" as Babbage called them. When their possible interpretation or application is considered, one may, of course, entertain various contexts and potential *designata*. But the move "up," by suggestion, from the subtraction of numbers and quantities to the symbolical formula $a - b$ on no account requires contemplating the performative extension of subtraction *within* the realm of number! What it does require is a willingness to regard the operation in radically pure syntactical abstraction. But that is a different and decidedly non-Lockean matter entirely.

Peacock's reasoning is weak, but the normative uneasiness that motivated him was strong. The only way algebra as he knew it could be preserved and justified intact was by setting it loose from arithmetic or, at the very least, setting loose those parts of it that allegedly defy arithmetic interpretation. But algebra as Peacock knew it constituted a logically connected whole that could not be broken down into independent numerical and nonnumerical components. His solution, therefore, was to retain symbolical algebra intact as a system, while breaking it down into *numerically interpretable* and noninterpretable components. To do so, however, amounted to rendering arithmetical algebra wholly subordinate to symbolical algebra and, contrary to his strong Cambridge-cum-Lockean bias, to stripping the numerical intuitions that ground arithmetic of their traditional constitutive role for algebra.

And thus, contrary to his own firm introductory statement, in order to preserve a constitutive role for arithmetical algebra, Peacock ended up reversing his own formulation by requiring interpretation, as it were, to *precede* rather than to follow "the operations of symbolical algebra." To formally ensure in advance that, when interpreted symbolical algebra would yield arithmetical algebra in its entirety, Peacock formulated the foundational principle that grounded his twofold algebraical system. He called it the "Principle of the Permanence of Equivalent Forms." "Equivalent forms" are algebraical identities. The principle of their "permanence" serves to guarantee their preservation each time they are "transported" across the arithmetical/symbolical divide. This requires separate principles for crossing back and forth. Peacock formulated them in the form of two "propositions":

> *Direct proposition*: Whatever form is algebraically equivalent to another when expressed in general symbols, must continue to be equivalent, whatever those symbols denote.
>
> *Converse proposition*: Whatever equivalent form is discoverable in arithmetical algebra considered as a science of suggestion, when the symbols are general in form, though specific in their value, will continue to be an equivalent form when the symbols are general in their nature as in their form. (198–99)[52]

The Converse Proposition serves formally to clinch the foundational role of arithmetical algebra for symbolical algebra in Peacock's dual system. Fully generalized and abstracted, their symbols and operations carefully dispossessed of all numerical connotation, each and every equation of the former is automatically forced upon the latter. This, of course, amounts to much more than "suggestion." "Determination" seems to be the proper term.[53] As noted above, although this might seem like the final step in a Lockean ascent of progressive abstraction from the most primitive notions of number and practices of numeration, the supposedly last move up from arithmetical to symbolical algebra is categorically different from all preceding steps. In moving up from counting to the "science of counting," from that to arithmetic, and from arithmetic to arithmetical algebra, numerical meaning, the *idea* of number or quantity, is not only preserved, but by keen self-reflection (as Frend would have it) or by keenly reflecting on the history of human counting and calculating (as would Peacock),[54] it is increasingly better articulated and increasingly more subtly conceived. On such a showing, the laws of arithmetical algebra *make explicit*, if you wish, an elaborately connected system of propositions deemed by reflection to be true *of* the idea of number or quantity. This no longer holds true of the relationship

Peacock proposes between arithmetical and symbolical algebra. The whole idea of symbolical algebra is that both its symbols and its operations are thoroughly stripped of all conceptual content, rendering it a purely formal system whose symbols, in and of themselves, no longer pertain to number or quantity, and whose formulas, in and of themselves, can, therefore, no longer be deemed true *of* anything. The only notion of truth to which they do pertain is what is referred to in modern parlance as truth by *coherence*. They articulate nothing and make nothing explicit. Peacock's symbolical algebra is obtained by radically generalizing generalized arithmetic—so radically as to totally sever the Lockean process by which his arithmetic project so forcefully set forth.

Symbolical algebra may, in principle, end up including operations and identities that resist arithmetical interpretation, but, unlike Babbage's pure analysis, Peacock's system prima facie disallows the reverse. Given the Converse Proposition, there can be no such thing as an exclusively arithmetical truth. All arithmetical truths automatically generate valid symbolical formulas. The move up from quantitative equality to formal identity, states the Converse Proposition, is purely a matter of abstraction and, in principle, can harbor no exceptions. In Babbage's system, to recall, pure symbolical identity is significantly more restrictive for pure analysis than quantitative equality is for arithmetic. Accordingly, as we have seen, Babbage banished all talk of infinite series from the former, although he regarded them as perfectly legitimate in the latter. This type of move is not merely absent, but is quite impossible in Peacock's algebra, in which, for lack of a separate standard, all valid arithmetical equations necessarily end up, suitably abstracted, as symbolical identities.[55]

The Direct Proposition—which supposedly imposes upon arithmetical algebra every symbolical identity deemed arithmetically interpretable—might appear to redress the balance by working in the opposite direction. Peacock, after all, promised that symbolical algebra should be "constructed upon its own rules" independently of, and prior to, its application (back down) to arithmetical algebra, or any other of its applications. But the sense of independence, logical primacy, and autonomy created for symbolical algebra by the Direct Proposition and his preemptive rhetoric turns out largely to be an illusion. Throughout the *Treatise* the Direct Proposition is applied only to low-level results—never at the level of fundamental principle or rule of operation. All the fundamental principles of symbolical algebra, its definitions and rules of operation, are, without exception, funneled up from arithmetical algebra courtesy of the Converse Proposition.

His treatment of exponents (201 and following) is typical. The operation is defined for arithmetical algebra in the usual manner as shorthand for repeated self-multiplication:

$$\text{def:} a^n = \underbrace{a \cdot a \cdot a \cdot a \ldots a}_{\text{n times}}$$

from which are derived the three well-known arithmetical "equivalent forms" for m and n whole and positive:

$$a^m \cdot a^n = a^{m+n}$$
$$\frac{a^m}{a^n} = a^{m-n}$$
$$(a^m)^n = a^{m \cdot n}$$

The three formulas are then generalized by means of the Converse Proposition and rendered bona fide "equivalent forms" of symbolical algebra. No longer constrained, m and n can then be manipulated freely to yield such symbolical identities as:

$$a^{m/n} = \left(\sqrt[n]{a}\right)^m$$

which, in turn, are interpreted back down into arithmetical algebra, by means of the Direct Proposition, to yield such equations as:

$$a^{1/2} = \sqrt[2]{a}$$

In doing so, the original definition is transcended. Such equations defy the arithmetical definition of a^n, and are, therefore, unobtainable, of themselves, within arithmetical algebra. Nonetheless, they do not represent mere meaningless notational conventions. Once obtained, they are allowed to enter the three arithmetical formulas from which fractional exponents were originally barred. In doing so, the original stipulations requiring m and n to be whole and positive are, in effect, canceled, and from now on the three formula are permitted to apply *within arithmetical algebra* to all rational m and n, even though the chain of logical entailment from the original definition has been severed!

To a great extent this brief example summarizes the entire enterprise of the *Treatise*. In Peacock's theory there are simply no independent rules of symbolical algebra. What he misleadingly dubs "suggestion" turns out to be a systematic and exhaustive, two-tier process of *abstraction* of arithmetic by which, in its second stage, the entire system of symbolical algebra is *generated*. Without exception, all deductive points of commencement within symbolical algebra have their origin in arithmetical algebra, systematically offered up by virtue of the Converse Proposition. Only lower-level, secondary symbolical results (e.g., $a^{m/n} = \left(\sqrt[n]{a}\right)^m$) are interpreted back down and imposed upon arithmetical algebra care of the Direct Proposition.

The *Treatise*, I have argued, was the work of an ambivalated Lockean meta-mathematical "theorist," forced reluctantly to compromise his principles, yet unable to turn his back on them completely. In his desire to carve a foundational role for the science of number, Peacock, in effect, fashioned symbolical algebra as a subsidiary of arithmetical algebra rather than the other way round (much rhetoric to the contrary notwithstanding). Though unconstrained, wider in scope, and laying no claim to either meaning or truth, symbolical algebra crucially remained a derivative (by maximal abstraction) of arithmetical algebra and, hence, a far cry from the type of truly independent, freestanding formal system envisaged and partly developed by Babbage before him, and later suggested by Gregory. Babbage's pure analysis purported to provide the *syntactical* basis for all mathematics, the fountainhead that nourishes and lends structure to all fields of mathematical analysis. In the case of exponents, to stay with our current example, Babbage's point of departure would not at all be, as in Peacock, the numerical definition of repeated multiplication (on the side of arithmetical algebra), but the three purely formal rules of operation:

$$a^p \cdot a^q = a^{p+q}, \frac{a^p}{a^q} = a^{p-q}, \left(a^p\right)^q = a^{p \cdot q}$$ [56]

jointly construed (on the side of symbolical algebra) as the purely formal and "official" *definition* of the operation. Remaining in the realm of pure analysis, all algebraic results are formally obtained from the tripartite definition *prior to* their application to number. In Babbage's system symbolical algebra is the foundation and arithmetical algebra the byproduct—the former, deductively complete, clean, purely syntactical, and well founded; the latter, meaningful, partly parasitic, necessarily incomplete, messy. Peacock wanted it the other way around, and in all probability would have liked both of his algebras to boast completeness. However, instead of enjoying the best of both worlds, the system he proposed in 1830 ended up plagued by the sum of their worst.

Wholly parasitic on arithmetical algebra, Peacock's symbolical algebra adds up to little more than an elaborate, disconnected set of exercises in exploring the consequences of ignoring the former's conceptual basis. Results are first obtained in arithmetical algebra, by virtue of the foundational definitions and semantic principles of that system, and are only then emptied of content and further manipulated. It is not an exercise in seriously determining algebra's syntactical infrastructure, as was Babbage's.[57] Peacock's *Treatise* did nothing of the sort. What, then, was the point of his elaborate twofold system? The obvious answer is that by setting arithmetical algebra loose from its numerical moorings, one was able to prove a large number of otherwise unobtainable results (e.g., $a^{p/q}=\left(\sqrt[q]{a}\right)^p$) that when applied back down, enriched arithmetical algebra where the likes of Masers and Frend had so dismally failed (e.g., $a^{1/2}=\sqrt[2]{a}$).

As we have seen, reasoning up from arithmetical to symbolical algebra by means of the Converse Proposition deviates considerably from the Lockean framework to which Peacock was so firmly committed. This represents the ambivalated Peacock's normatively hybrid concession to Babbage's radically formalist challenge. However, the price he had to pay for maintaining the uneasy coupling was enormous. Reasoning up from arithmetical to symbolical algebra, courtesy of the Converse Proposition, is at least logically consistent. But the same cannot be said of the way the *Treatise* reasons in the opposite direction, courtesy of the Direct Proposition. Each time a symbolical result, unobtainable of itself in arithmetical algebra, is applied by means of the Direct Proposition back down to arithmetical algebra, Berkeley's fallacy of "shifting of the hypothesis" is inevitably committed! Obtaining $a^{p/q}=\left(\sqrt[q]{a}\right)^p$ on the assumption that p and q are vacuous symbols and applying it as a general result to a system in which a^b is defined exclusively for bs whole and positive is the same as manipulating an expression under the assumption that Δx is something and, after dividing by it, deciding in mid-proof that it has vanished![58] Peacock makes such moves all the time, enthusiastically exploring the consequences for arithmetical algebra of generalizing the indexicals of indefinite series, the coefficients of cyclical functions, the orders of differentiation, and even factorials.[59] To use the Direct Proposition to supplement arithmetical algebra with results that, of themselves, fly in the face of its basic principles, strictures, and definitions is at best arbitrary, and at worst fallacious. There is nothing inherently wrong in developing the two algebras proposed by the *Treatise*. Nor is there anything inherently wrong in creating a symbolical algebra by abstracting generalized arithmetic. But if the point of doing so is to somehow improve arithmetical algebra by enriching it with otherwise unobtainable results, then Peacock's system is quite inadmissible. A result cannot be considered *true of* the realm of numbers if, to obtain it, one is forced first to assume that it is not.

Peacock never openly acknowledged the existence of such a problem. Nor, to the best of my knowledge, did any of his friends or critics, except perhaps for Whewell, who, as we shall see in the chapter to come, even during the brief period in which he eagerly endorsed the *Treatise*, shows clear signs of discomfort with the Direct Proposition. Whewell nowhere explains what he found wrong with Peacock's account of the relationship between arithmetical and symbolical algebra, and I have found no written evidence of the two men discussing it, although, as close friends and Fellows of the same college, they must have talked long and hard about all matters algebraic. Whatever the nature of these conversations, or the extent to which Whewell's criticism might have borne on the further development of Peacock's views (Peacock certainly did not change anything between the *Treatise* of 1830 and BAAS Report of 1833), it is clear that eventually Peacock would also see fit to do away with the Direct Proposition—which, as we shall see, is carefully omitted from the second edition of the *Treatise*, published during the early 1840s.

First, though, let us turn to how the original *Treatise* was received.

CHAPTER SEVEN

Peacock's Impact

INTRODUCTION

For all its deficiencies, the *Treatise* was an extraordinarily influential work, in relation to which Whewell, DeMorgan, William Rowan Hamilton, D. F. Gregory, and others were forced to formulate and reformulate their own positions, and did so with explicit reference to Peacock.[1] The historical significance of Peacock's work lies primarily in the problems it raised for its various readers, eliciting responses that varied as much as its readers' prior convictions. For a theoretician at heart, such as Hamilton, Peacock's renunciation of a truth relation for symbolical algebra presented the gravest challenge. For the more philologically inclined, such as Gregory and the young DeMorgan, the lack of an autonomous and coherent set of foundational principles was perceived as its most conspicuous flaw. Pronouncements to the contrary notwithstanding,[2] no one was able to accept Peacock's initial solution to either problem, namely, to view arithmetical algebra in the dubious Janus-faced role of a "science of suggestion" that stood in for the truth, meaning, and independent foundation (as the case may be) lacking in its symbolical counterpart and at the same time, as its major beneficiary, enjoyed symbolical algebra's symbolical fruits in a manner that gravely subverted that very truth and meaning![3]

It is well beyond the scope of the present study to characterize and compare the considerably less hybridic accounts of algebra proposed in the works of Whewell, Gregory, DeMorgan, and Hamilton in any detail. My focus here will be on how each of them read Peacock's *Treatise* and the ambivalating effect it had on their work. Lukas Verburgt's excellent recent work on the immediate responses to the *Treatise*[4] nicely complements the account presented here in this regard, despite our interestingly different historiographical approaches and differently nuanced readings of Peacock, Gregory, and Hamilton.[5]

DEMORGAN'S CHANGE OF HEART

It took DeMorgan the best part of five years to digest the *Treatise* sufficiently to write a review, and much longer to form a consistent and better thought-out theory of his own.[6] In his later work DeMorgan endorsed Peacock's basic differentiation between symbolical algebra and its models, but dispensed with the "suggestive" role Peacock assigned to arithmetic. He attempted to restore for algebra the logical hierarchy of a formal system and its models, grounding the former on a set of axioms and relegating truth and meaning exclusively to the latter. "When we wish to give the idea of symbolical algebra," he wrote in 1840,

> we ask, firstly, what symbols shall be used (*without reference to meaning*); next, what shall be the laws under which such symbols are to be operated upon; the deduction of all subsequent consequences is again *an application of common logic*. Lastly, we explain the meanings which must [*sic*] be attached to the symbols, in order that they may have prototypes *of which the assigned laws of operation are true*.[7]

Two years later DeMorgan would be the first to propose a set of formal axioms for algebra.[8]

But even he would remain partially ambivalent, at least until the mid-1840s. In accepting the idea of an independently founded symbolical algebra, he certainly advanced considerably beyond Peacock's initial rude divide. Nonetheless, he recoiled from the idea of an autonomous "technical algebra" (as he termed "the art of using symbols under regulations which . . . are prescribed as the definition of the symbols"),[9] which he felt should not be considered a science, and bore no resemblance to what mathematicians actually did. To return to the well-known passage cited in the previous chapter, DeMorgan likened one "who makes the transformations of algebra by the defined laws of operation only . . . to one who puts a dissected map together by the backs of the pieces alone; whereas the person who looks at the front, and uses his knowledge of geography to help, more resembles the investigator and mathematician."[10]

DeMorgan had started out, as Joan Richards has shown, deeply committed to the form of Lockeanism she terms "Pauline."[11] The effect of Peacock's *Treatise* on his thinking was, therefore, profound. Peacock's work convinced him, more than it had convinced its author, of the primacy of symbolical algebra as the foundation of mathematics. But, like Peacock, though not to the same extent, DeMorgan continued to dither between the two incompatible frameworks: committed to symbolical algebra, yet unable to fully comprehend and endorse a Babbage-type formalism as a viable single meta-mathematical option.[12]

When William Rowan Hamilton first read the *Treatise*, he was as deeply committed as the young DeMorgan had been to a decidedly nonformalist understanding of algebra "as a Science properly so called; strict, pure, and independent" that viewed its axioms as harboring an immanent truth relation, as "deduced from its own *intuitive* principles; and thus not less an object of priori contemplation than Geometry."[13] But unlike DeMorgan, Hamilton resolutely opposed conceding Peacock's contention for the need for symbolical algebra. In fact, he appears to have mistaken Peacock's hybrid account for a position much closer to that of Babbage. Like DeMorgan, it would also take Hamilton five years to properly respond to the *Treatise*. But despite his rejection of Peacock's main suggestion, the impact of the *Treatise* on his work would be even more profound—life-changing, one might say.

HAMILTON'S "ALGEBRA AS THE SCIENCE OF PURE TIME"

If DeMorgan initially responded to reading the *Treatise* by reviewing the work, Hamilton responded to it by dropping everything else in order to devote himself fully to the foundations of algebra and to formulating a fully fledged, radically different, and fiercely original approach of his own. Upon learning, in July 1835, that Peacock was unable to attend the August meeting of the British Association for the Advancement of Science in Dublin, Hamilton wrote him as follows:

> Besides the pleasure which I expected from even merely seeing you again, I wished to talk to you about Algebra. I have taken some pains to understand your views, & almost persuaded myself that I do. They interest me much, & I hope to study them more; but you have perhaps perceived enough of the turn of my mind in metaphysics to expect that I should rather admire them than adopt them. So far, indeed, have I gone in the opposite direction, that wishing lately to publish some sketch of my own views on the subject, I found myself forced back on an old thought

> of mine, & have nearly printed an essay on Algebra as the Science of Pure Time. The introductory remarks, however, (without expressly mentioning you)[14] recognize the existence (I might say the necessity) of a Philological School of Algebra, in which school you perhaps, have been the most bold and consistent teacher."[15]

The essay to which Hamilton's letter refers was embarked on in 1832, completed in 1835, and published in the *Transactions of the Royal Irish Academy* of 1837.[16] It would take Hamilton until 1850 to complete his ambitious project,[17] but by the time it was completed, although his approach to algebra would remain unchanged, his attitude to Peacock's approach would curiously and considerably mellow. In the 1837 essay, Hamilton formulated the threefold typology of algebraists (from which I quoted at length in chapter 5), rejecting the "philological" formalism of Peacock's symbolical algebra (without naming Peacock) in favor of a thoroughly "theoretical" approach. However, Hamilton's point of departure, unlike that of the young Peacock, was not Lockean. His orientation, as Whewell's had by then become, was primarily Kantian—though in Hamilton's case, his Kant was strongly mediated by Coleridge.[18] Algebra, he insisted, was a science, not a grammar—a body of truths true not of number or quantity, but of the ordinal characteristics of "pure time." And as in Kant (as opposed to Locke) the idea and understanding of time, like those of space, were, for Hamilton, not gleaned off the delivery of the senses but, in being constitutive of sensual experience, were considered necessarily prior to and independent of all experience.

In the preface to his *Lectures on Quaternions*, published in 1853, Hamilton returns both to the tripartite taxonomy of algebraists presented in the 1837 paper, and to the idea of algebra as the science of pure time. From the start it was clear to him that an algebra founded on the ordinal characteristics of single moments in time was still incapable of accommodating imaginaries. The 1837 paper proposed developing algebra as a theory of "moment couples," which Hamilton then attempted unsuccessfully to extend to a theory of triads.[19] In moving up from triads or triples to quaternions and beyond, nothing of the intuitive ordinality of time is lost. And yet, Hamilton claimed explicitly that the "method" or "calculus" of quaternions, as he called it,[20] brought him much closer to Peacock's two algebras approach than he had originally thought—as if the algebra of quaternions constitutes an algebra that "suggests" a purely formal system in the same way as Peacock's arithmetical algebra stands to its symbolical counterpart. In a lengthy footnote to the preface of his *Lectures on Quaternions*, Hamilton describes his project in explicitly Peacockian terms, very differently than he had done in 1837:

> I am very willing to believe that . . . I may have attended too little to the *symbolical* character of Algebra, as a language, or organized system of *signs*: and too much (in proportion) to what I have been accustomed to consider its scientific character, as a Doctrine analogous to Geometry, through the Kantian parallelism between the *intuitions* of Time and Space. . . . Yet I wish to state, that I do not despair of being able hereafter to shew that my own old views respecting Algebra, perhaps modified in some respects by subsequent thought and reading, are not fundamentally and irreconcileably opposed to the teaching of writers . . . as Drs. Ohm and Peacock. . . . The *quaternions* . . . and even the *biquaternions* . . . might have their laws explained, and their symbolical results interpreted, by comparisons to *sets of moments*, and by operations on *sets of steps* in time. Thus in the phraseology of Dr. Peacock, we should have a very wide "science of suggestion" (or rather, suggestive science) as our basis, on which to build up . . . a new structure of purely *symbolical generalization*, if the science of time were adopted, instead of merely Arithmetic, or (primarily) the doctrine of *integer number*.[21]

Seven years earlier, in October 1846, Hamilton conveyed his interesting change of heart regarding the *Treatise* in a letter to Peacock. "When I first read that work," he wrote,

> and indeed for a long time afterwards, it seemed to me . . . that the author designed to reduce algebra to a mere system of symbols, and *nothing more*; an affair of pothooks and hangers, of black strokes upon white paper, to be made according to a fixed but arbitrary set of rules: and I refused, in my own mind, to give the high name of Science to the results of such a system; as I should, even now, think it a stretch of courtesy . . . to speak of chess as a "science."

However, since then, he goes on to say, "[m]y views respecting the nature, extent and importance of symbolical science may have approximated gradually to yours," despite the fact that "I still look more and more habitually *beyond* the symbols," than the more avid formalists "would choose to do. To use DeMorgan's image, I turn up the fronts of the bits of the dissected map, and try to learn what countries they denote as well as how they fit together."[22]

The curious thing about Hamilton's reconsideration of Peacock's position is that it did *not* at all attest to a "Peacockian" shift in his own work. Moving up, in Hamilton's system, from moments to couplets, triplets, quaternions, or biquaternions of moments is not achieved by generalizing, or abstracting from earlier levels. Quaternions are not generalizations of, or in any sense entities more abstract than, couplets and triads. The process cannot be said, therefore, to culminate in a symbolical system progressively emptied of algebraic meaning. In Hamilton's version of the metaphor borrowed from DeMorgan, the blank fronts of the bits of the dissected map relate and mesh according to their form alone, but their backs contain the meanings they denote. Of themselves, the "equivalent forms" of Peacock's symbolical algebra denote nothing—"back" or "front"—they are purely formal identities.

Hamilton more than hints that his concessions to Peacock had to do with the developments of the latter's system since the *Treatise* and BAAS Report: due, as he puts it, to "my impression from our last interviews" that "you . . . do not only not undervalue interpretation, as indeed your work attests, but would even desire, if any changes were to be made, that the importance of such interpretation should be brought more prominently forward than it was by you, at least in your first work on algebra."[23] As we shall see, this goes to the heart of Peacock's own considerable reformulation of his algebra during the early 1840s, in the second edition of the work.

But then, maybe it does not. Perhaps DeMorgan's image of the dissected map should be taken more seriously, and be understood as pointing, in both his and Hamilton's case, to a more subtle form of endorsing Peacock's master idea. For one thing, as I remarked in passing, clever as it is, the metaphor is quite inappropriate as a description of Peacock's position in either of its editions. For Peacock, arithmetical and symbolical algebra were, and were to remain two distinct and separate, if related fields of mathematics. The only thing he significantly modified in the second edition was the nature of their relationship. Throughout his career, the two algebras themselves remain intact and freestanding enterprises, representing in their very separateness the stark irreconcilability of his two jarring commitments. The appropriate image for Peacock's system would be that of *two* dissected surfaces, one painted and one blank. DeMorgan's image of two-sided pieces of one dissected map or jigsaw puzzle suggests a very different kind of hybridity—not two systems pertaining each to different forms of mathematical meaning and truth, but one system comprising "two-sided" elements, each pertaining at once to a *dual*, or *twofold*, form of meaning and truth! Could DeMorgan, and especially the later Hamilton, be read as sympathetically reading Peacock in this way, namely, as proposing a unitary view of algebra as a single system perfected and developed with a view to attaining both formal *and* nonformal objectives? Such an idea would certainly resonate nicely with the nonexclusivity of Hamilton's initial threefold taxonomy and its realization in quaternions, as well as the algebras on which he and DeMorgan were working.

If so, then Hamilton's changing attitude toward Peacock's algebra emerges as the result of two consecutive misreadings. In the first instance, he mistook the *Treatise* (not unlike D. F. Gregory and several of his latter-day readers) to be advocating an ultimately formalist (philological) approach to algebra similar to Babbage's as the culmination point of a process of "reasoning up" from arithmetic—a position Hamilton firmly opposed, and in opposition to which he embarked on his "Algebra as the Science of Pure Time." At a later stage, moved to reconsider perhaps by DeMorgan's metaphor, Hamilton again mistakenly came to see Peacock as advocating as the culmination point of his algebra project a Janus-faced, yet *unitary* view of algebra, perceived as a *single* connected system of equivalent forms that boasted at once both arithmetical and symbolical perfection—as if, even when, as in the second edition, the two algebras were relegated to separate volumes, they were still meant to be read as two different renditions of the *same thing*.

Such a view, even if unjustified as a viable interpretation of Peacock, also resonates extremely well with the way in which, I shall argue, Whewell in his *Philosophy of the Inductive Sciences* of 1840 would eventually try to accommodate Hamilton's own, far more "Peacockean" insistence on there being two quite separate dynamical sciences. By 1840, Hamilton's two sciences of 1834 had, in Whewell's system, become one: a single unitary, yet Janus-faced body of knowledge cultivated at once as a well-formed, deductively structured body of theory, and as a perfect "colligator" of empirical fact. On my reading of his philosophy, the propositions making up a true Whewellian inductive science are perfectly analogous to the two-sided pieces of DeMorgan's dissected map.

If this is true, then DeMorgan's, Hamilton's, and Whewell's thinking during the 1840s, undertaken originally at least partly in response to Peacock, can be seen to demonstrate an interestingly more subtle form of hybridity than Peacock himself had envisaged. As opposed to his external coupling of two algebras, the hybridity their writings exhibited was *internal* to the systems they cultivated. In themselves, no less shaky, and no less jarring, perhaps, but far more subtly accommodated within a supposedly single, unified system.

D. F. GREGORY'S "ON THE REAL NATURE OF SYMBOLICAL ALGEBRA" (1840)

If Hamilton can be seen as reacting to the reformulation of the *Treatise* undertaken by Peacock during the early 1840s, D. F. Gregory's "On the Real Nature of Symbolical Algebra," written seven or eight years earlier,[24] with its grave misreading of Peacock's original position, may well have been instrumental in prompting Peacock to reconsider.

Gregory's paper is, to the best of my knowledge, the first published work of meta-mathematics to systematically conceive of symbolical algebra, as did Babbage's unpublished "Philosophy of Analysis" of almost twenty years earlier, as a wholly formal calculus of operations established on its own principles, independently of and firmly "distinguished from the various branches of analysis that come under its dominion."[25]

Rather than view the formal resemblance of, say, $a^m \cdot a^n = a^{m+n}$ and $(d^n/dx^n)(d^m u/dx^m) = d^{m+n}u/dx^{n+m}$ as a mere structural analogy, Gregory takes them as pertaining to a shared formal law of operation by which the two lower-level operations are governed. In fact, as he points out in the following example, structural analogy and subsumption under the same operational law are not necessarily synonymous. The operations of addition and subtraction in arithmetic and those of transferring a point through a circle's entire and semi-circumference in geometry are both subject to the same laws of operation defined quite generally with respect to two classes of operations F and f thus:

(1) $FF(a) = F(a)$	(2) $ff(a) = F(a)$
(3) $Ff(a) = f(a)$	(4) $fF(a) = f(a)$

Nonetheless, he writes, "we see clearly . . . that there is no real analogy between the nature of the operations . . . in arithmetic and geometry, as is generally supposed to be the case." Nothing in the nature of arithmetical addition involves anything akin to coming full circle in geometry, and vice versa. Indeed, says Gregory, "the two operations cannot even be said to be opposed to each other in the latter science as they are generally said to be" (3–4).

Gregory conceived of symbolical algebra as an independent science:

> the science which treats of the combination of operations defined not by their nature, that is, by what they are or what they do, but by the laws of combination to which they are subject. (2)

He was clearly deeply influenced by Peacock, whom he credits for being, to the best of his knowledge, "the only writer in this country who has attempted to write a system of algebra founded on a consideration of general principles"(1), which "have been in many cases suggested . . . by the laws of the known operations of number"(2), for like his older colleague he goes on—unfoundedly, if more warily—to formulate a strong version of the *Treatise*'s Direct Proposition. The step from arithmetical to symbolical algebra is achieved, he submits, by "leaving out of view the nature of the operations that the symbols we use represent" while attending exclusively to their laws of combination, between which "we are able to prove certain relations" that "when expressed between the symbols, are called algebraic theorems" (2).

The step "up" from arithmetical to symbolical algebra is fine, as is its further development as an independent science of operations.[26] It is the step back down from symbolical algebra to the "different classes of operations," described by Gregory as "*included in* the general case,"[27] that is problematic. On the one hand, he states that insofar as it can be shown that "any operations in any science are subject to the same laws of combination . . . the theorems are true of these as included in the general case"—which is Gregory's version of the *Treatise*'s Direct Proposition. It is different from Peacock's, however. Peacock is careful *not* to describe symbolical algebra as a generalization or interpretation of its arithmetical counterpart. In the *Treatise*, the Direct and Converse Propositions are formulated not as entailed by, but as *definitive of* the relationship between the two algebras. Unlike Peacock, and for that matter also contrary to Babbage, with whose work on pure analysis he was, in all probability, unacquainted, Gregory appears to construe symbolical algebra from the start as a generalization of arithmetic, which renders his version of the Direct Proposition a *necessary consequence* of the relation between the two, claiming that "whatever can be proved of the class generally, is *necessarily true* of all the operations included under it."

On the other hand, it is equally clear that the particular meanings that the operations in question receive in their lower-level domains may significantly limit the extent to which the theorems of symbolical algebra may at all be applied to them. "What they are, and what they do," in their original domains, as Gregory has it, may very well render a theorem "inapplicable to that branch of science," despite being necessarily true of it! And it is here that Gregory's and Peacock's projects subtly diverge, in a manner that, as we have seen, cripples the latter.

That the nature of the *applicandum* is liable to constrain the *applicans* is not problematic in itself. An algebraic problem that involves calculating, say, the number of people in a building will not admit solutions that are not positive and whole, regardless of whether they are validly yielded by the equations. Sometimes algebra's overdetermination is liable to expose a constraint as unjustly narrow, and teach us something unexpected about the *explanans*, as in the famous story of Dirac's discovery of the positron.[28] But such cases are rare, and bracketing mere oversight, the constraints in question are largely predetermined by the nature of the entities and operations of the domain in question. If there is any truth in my reading of Peacock, the ultimate aim of his two algebras project was *not* to establish algebra anew as symbolical algebra, as was Babbage's intention, but to use symbolical algebra to enrich arithmetical algebra by transcending the latter's "Lockean" constraints courtesy of the Direct Proposition!

In this respect, Gregory's project was significantly different, steering (unaware) much closer to Babbage than to Peacock. Its aim was to develop symbolical algebra as an independent calculus of operations, regardless of whether, how, or by which particular lower-level domain its various definitions were initially suggested. And as far as I can tell, it was also a project wholly unhampered by Lockean hangups. The precious little Gregory left us in writing does not make the meta-mathematical point explicitly, but his discussion of particular cases wears it on its sleeve.

Take, for example, still again Peacock's account of a^m, which, as we have seen, sets forth from the arithmetical definition of repeated self-multiplication, yielding $a^m \cdot a^n = a^{m+n}$, $a^m/a^n = a^{m-n}$, and $(a^m)^n = a^{m \cdot n}$ for any quantity a, and positive integers m and n. Courtesy of the Converse Proposition, the three arithmetical formulas are then rendered equivalent forms of symbolical algebra by stripping a and m of all meaning and reversing the arrow of definition. In symbolical algebra a^m is *defined* by means of the three operational laws rather than vice versa. Unconstrained, the three laws entail several additional "theorems," such as $a^{m/n} = \left(\sqrt[n]{a}\right)^m$ which are then applied back down to arithmetical algebra, by virtue of the Direct Proposition, in order to enrich the arithmetical scope of a^m by allowing for all manner of powers neither positive nor integral. Of course, all talk of negative, zero, and fractional powers of a fly in the face of the original definition of self-multiplication.

Gregory's account is different. Commencing from symbolical algebra, he asks us to consider the existence of a "genus of operations" f_i subject in general to the following laws:

$$f_m(a) \cdot f_n(a) = f_{m+n} \qquad (1)$$

$$f_m f_n(a) = f_{mn} \qquad (2)$$

If "the form of f" is further defined by making $f_1(a) = a$, and m and n are taken to be integer numbers, we are then in the realm of what Gregory dubs "index operations"—"operations which are represented in arithmetical algebra by a numerical index" (5). Thus, he explains,

> if m and n be integers, and the operation a^m be used to denote that the operation a has been repeated m times, then, as we know, $a^m \cdot a^n = a^{m+n}$, $(a^m)^n = a^{mn}$

Gregory's next point, however, nicely exemplifies how and where he parts company with Peacock. "We have now to consider," he continues,

> whether we can find any other operations besides that of repetition which shall be subject to the laws we have laid down. If we suppose that m and n are fractional instead of integer, we easily deduce from our definition that the notation $a^{p/q}$ is equivalent to the arithmetical operation of extracting the qth root of the pth power of a, . . . and we find without difficulty, from the law of combination, that a^{-m} indicates the inverse operation of a^m, whatever operation a may be.

For Gregory these are examples decidedly *not* of arithmetical enrichment by virtue of moves made in symbolical algebra, but of the search for their analogues and manifestations in *existing* arithmetical operations that can be shown to abide by the symbolical laws in question. Gregory considers symbolical algebra as a branch of mathematics separate from and independent of the "various branches of analysis which come under its dominion"(1) and irrelevant to their further development—including arithmetical algebra, which he takes to be fully developed prior to seeking for arithmetical instances of symbolical laws. And yet, he describes his project as taking a view of the subject "more general" than that of Peacock (1), but not as deviating from Peacock's initial conception in any way.

Whether Gregory's response owed to a simple misreading of Peacock's *Treatise* and BAAS Report of 1834, or to a back-beaming of the second thoughts that Peacock might have been experiencing during the late 1830s, to which Gregory was likely to be exposed at Trinity, is hard to say. Be that as it may, Gregory's paper, though presented as continuing the *Treatise*'s project, proposed an innovative, yet modest, decidedly *non*hybridic account of symbolical algebra as a new calculus of operations, which purported neither to be foundational, nor to supply the key to arithmetical algebra's further development. Unlike the *Treatise*, it was not a work of meta-mathematics, nor one motivated by meta-mathematical concerns. But by presenting itself as a continuation or completion of Peacock's project, it served dramatically to obfuscate and conceal for his readers the profound ambivalence to which the *Treatise* owed its origin and structure, and to place on the agenda for the first time the seemingly uncomplicated outlines of a truly formalist account of algebra. I say "seemingly," because unlike Peacock, Gregory wrote as if arithmetical algebra could be fully developed independently of symbolical algebra in ways capable of transcending the constrictions of generalized arithmetic, which, as Peacock knew all too well, it could not. And thus, following closely in his footsteps, Gregory was so taken with his brilliant tutor's innovative redescription of algebra that his development of it managed unwittingly not only to misinterpret his solution, but to hide from view the problems it harbored that it sought to solve.

WHEWELL'S FAILED PHILOSOPHY OF HIGHER MATHEMATICS

Whewell was quickest to respond, but was ultimately less successful than DeMorgan, Hamilton, or Gregory in constructing an account of higher mathematics of his own. He certainly tried, and on a substantial scale in a broadly conceived attempt during the early 1830s, to incorporate an appreciative account of Peacock's two-tier approach to algebra as the basis for a comprehensive philosophy of mathematics. Whewell's one-hundred-page manuscript, entitled "Philosophy of the Pure Sciences,"[29] was abandoned, however, sometime after he penned the last entry, dated 1833. The text by the same name that was later published as book 2 of the *Philosophy of the Inductive Sciences* fell significantly short of the fully fledged philosophy of mathematics it had originally set out to formulate. The published version of 1840 described geometry and algebra and the nature of geometrical and algebraic truth along nonformalist, Kantian lines, describing them as the pure sciences of the "forms of intuition" of space and time.[30] But, as in Kant's work, it never came close to addressing the problem of viably accommodating modern algebra in its full richness. Whewell was probably aware of Hamilton's 1837 attempt to formulate such an algebra, but shows no signs of having contemplated going in the same direction. He seems to have steered too close to Peacock to appreciate Hamilton's countermove, and upon realizing his failure to accommodate the *Treatise* within the semi-Kantian framework he was developing, he lost either interest or hope.

The story of Whewell's aborted attempt to incorporate Peacock's algebra within his developing philosophy is a telling one, if only because it owed to a different tension harbored by the Analytical Society's original program, with which Whewell, at the time, was in the last stages of coming to grips. Although the society's founders enthusiastically embraced Lagrange's version of the calculus, none of them followed his lead into physics, at least not initially.[31] Lagrange's analytical mechanics was as formalist as his calculus: nonrealistic, noncausal, and subsequently nonexplanatory. Here, as we have seen, the young Cambridge reformers' native commitment to a Baconian view of the natural sciences remained firmly intact—inductivist, reductionist, realist, causal, and explanatory.[32]

In physics, as Smith and Wise rightly observe, it was Laplace rather than Lagrange whom the young "analyticals" took as their initial model.[33] What Smith and Wise fail to point out is the extent to which a compounded Lagrangian-Laplacian (or Lagrangian-Baconian) position implied by the society's program was inherently problematic in unwittingly implying a new and strained hybridic philosophy of mathematized science: Lagrangian (formalist, "philological") in its view of mathematics, and Baconian or Laplacian (realist, explanatory, "scientific") in its view of the study of nature.[34]

During the 1820s and early 1830s Whewell was the only one of the young reformers to attempt systematically to come to philosophical terms with this particular tension. I have told elsewhere the story of his awakening to it and of his several unsuccessful attempts to solve it before the mid-1830s.[35] By the early 1830s, Whewell had all but renounced his early commitments to Lagrange's calculus, Locke's empiricism, and Baconian induction, in favor of the partially Kantian yet realist position that would reach completion by 1840.

His problem with the *Treatise* was hence at this point similar to Hamilton's. On the one hand, he had to agree with Peacock that a viable philosophy of science at its best required a richer account of algebra than generalized arithmetic.[36] Peacock's symbolical algebra certainly fit the bill in that respect, but Whewell's newly found Kantianism could not allow it to be considered a merely formal system. In its final form, Whewell's philosophy would portray the conceptual component of any science, mathematized or not, as both synthetic and a priori, although he refrained from using Kant's famous terminology in his published and unpublished work.[37] His conception of it's a-priority diverted from Kant's, as we shall see, but its synthetic nature remained party-line Kant, and as such nonformalist in principle. He considered the axioms and proven propositions of any such system—all of mathematics included—to be true, not merely by definition, but of what he would later term the Fundamental Idea that governed it, describing the axioms and propositions of elementary geometry as true of the Fundamental Idea of space; those of arithmetical algebra, of that of time; those of the calculus ("higher geometry," as he dubbed it), of that of limit; those of dynamics, of that of force; and so forth. Whewell probably lacked the ability to *develop* such an algebra from scratch. He sought instead to reinterpret Peacock's symbolical algebra in nonformalist terms.

By December 1833 he had penned a fairly full draft of the first part of his planned *Philosophy of the Inductive Sciences*: a bulky hundred-page manuscript entitled "The Philosophy of the Pure Sciences." But by the time the *Philosophy* was published in 1840, the crucial section on the foundations of algebra had been largely suppressed, and the order of presentation altered to the extent that the text could no longer be considered a fully fledged philosophy of mathematics. Why?[38]

The 1833 draft comprises two large chapters entitled respectively "Of the Conditions of Perception and Their Consequences, or Of the Foundations of the Pure Sciences" (tellingly dithering between "Pure" and "Formal")[39] and "Of Aptitudes for the Pure Sciences, or Of Mathematical Talent." The latter is of interest, but of little relevance to his philosophy of mathematics, and to the best of my knowledge was never published. The former takes up most of the work and comprises fifteen sections. The first section, entitled "On the Necessary Forms under Which the Impressions of the Senses Are Conceived," is a sketchy forerunner of his notion of Fundamental Ideas and theory of the fundamental antithesis of philosophy, which would be fully developed only later. Sections 2–4 deal with what he would later term the Fundamental Idea of space, its basic features, the foundations of elementary geometry they generate, and the "impressions of the senses" from which, to borrow Kant's terminology, the idea of space is transcendentally deduced. Sections 5–8 treat in similar manner the Fundamental Idea of time and the idea of number it entails as furnishing the foundations of "Theoretical Arithmetic"—that is, arithmetical algebra devoid of negatives and imaginaries. All of this, appropriately edited and slightly differently structured, was later reproduced in the *Philosophy*. The two succeeding sections, "On Proportion" and "On Similarity of Geometrical Figures," were omitted in the *Philosophy*, but are of little consequence.

Section 11, entitled "On the Foundations of the Higher Geometry," is the first of four sections devoted to higher mathematics. Unlike elementary geometry, higher geometry deals with nonrectangular figures and therefore, according to Whewell, requires the introduction of a new Fundamental Idea in addition to that of space—namely, that of limit. This short section does little more than explain by example the "up to" axiom (the so-called continuity principle that "what is true up to the limit is true at the limit").[40] Finally, after "establishing the foundations" of arithmetic and geometry, and presenting each as a body of knowledge true of its foundational idea, Whewell turns in section 12 to consider in similar fashion the foundations of algebra, which was to prove the high point of his "philosophy of the pure sciences" project. (In section 13 algebra thus construed is subsequently applied back down to limits to found "the Higher Analysis in Mathematics," which, in turn, is applied in section 14 to displacements in order to found "the Pure Science of Motion.")[41] The idea was to describe algebra as what might be termed a meta-pure science—a maximally abstracted propositional scheme designed to apply to the lower-level pure sciences, "colligating" them in much the same way as, in his account of inductive science, the pure component of a well-formed inductive science applies to the facts.

> The distinguishing process of Algebra is the representation both of quantity and of operations upon quantity by symbols . . . we may represent two lines by a and b; a^2 may represent the square of a; ab may represent the product of the two lines, and $a^2 + ab + a(a + b)$ may represent a certain relation which can be proved arithmetically in any particular case *but of which this representation can also be shewn to be necessarily derivable from the elementary rules of such representations in general.* And in such deductions we do not refer to or think of the squares and products themselves, but only of the rules of combination which are assumed to prevail among the symbols.
>
> Algebra therefore takes for granted in the first place *the notion of symbolic representation,* that is of a representation in which one thing stands for another, in such a way that what is assumed of the one is somehow denoted by the other.[42]

Whewell considered the propositions of a pure science to be necessarily true (in Whewell's peculiar sense of the term) of the Fundamental Idea that governed it (as opposed to the colligatory truth or fit of the same propositions in relation to their *interpretanda* or *applicanda,* a twofold notion of truth that will be central to Whewell's philosophical system).[43] The question is, what is the governing idea of which the propositions of higher algebra could be said to be necessarily true? Whewell had adopted Peacock's basic division between arithmetical and symbolical algebra. Within Peacock's own framework the question was meaningless. Peacock's symbolical algebra, of itself, is true of nothing. What makes the 1833 draft truly fascinating is how it attempts to square the circle by salvaging a robust notion of truth for an admittedly thoroughly formal system by taking its Fundamental Idea to be that of absolute, or perfect, generality itself!

> In Algebra . . . the symbols are entirely arbitrary. . . . *This absolute generality of representation is the fundamental principle of all algebra, that is of all mathematical symbolical reasoning*; and it carries us . . . through all the possible cases of the science which no other principle will do.[44]

"[T]he above principle," he writes further down, "imposing the rule of perfect generality . . . is the only imperative and fundamental principle in the science."[45]

Such an option must have seemed plausible to Whewell in 1833, but as his Kantianism deepened, he seems to have realized its futility—that the idea of "absolute generality of representation" lacked in conceptual content to be something of which algebraic formulas could be considered true or false, in the same way that those of arithmetic are beholden to the idea of time or number.[46] Whewell's 1840 *Philosophy of the Inductive Sciences* described the ideal component of an inductive science as what lends all meaning to the body of sensual matter it "colligates," *conferring* upon it its entire conceptual content. Considered as a symbolic system generated by the idea of absolute generality, higher algebra would seem incapable of performing such a function. At best it could be said to formally *represent* bodies of lower-level arithmetical and geometrical truths, but of itself seems quite incapable of lending *them* any new meaning.

And yet, notes Whewell, we face:

> the paradox . . . that in a science like algebra, which professes to proceed altogether on a principle of representation there should occur symbols representing operations which cannot be performed & quantities which cannot be conceived, and that the use of such symbols should lead to right results.[47]

But what are these new and uninterpretable truths supposed to be true or "right" *of*? As noted, the form of Kantianism to which Whewell would very soon firmly subscribe left no room for mathematical truths true merely by definition, or by what would later be termed coherence. Truth within a Whewellian pure science is truth by correspondence, not to external reality, of course, but to features of the science's Fundamental Idea.[48] But if algebraic truth cannot be construed in relation to the Fundamental Idea of algebra, what are the grounds for ascribing meaning and truth to those parts of symbolic algebra that fail to "correspond" to low-level arithmetical truths? Whewell's answer in 1833 was to endorse Peacock's "principle of the permanence of equivalent forms," along with his bottom-up notion of "suggestion" coupled to an equally bottom-up notion of interpretation!

> [T]he meaning of a^n is assigned by considering the particular cases which it represents when n is whole, and by considering also that it must possess absolute generality of representation.[49]
>
> The same considerations would give us the meaning of all other symbols used in algebra. Operations in particular cases suggest the rule of the operation; *and the rule of the operation, expressed in symbols, defines the meaning of the symbols*; which thus obtain a generality of signification far wider than existed in the cases which were first represented. Thus the generality of algebraical symbols is a condition of their nature . . . and speaking with reference to the only sound principle of the science we must say, not that the general case of any symbolic representation is an extension of particular cases; but that particular cases are modes of interpreting the general symbols.[50]

When Whewell wrote these lines he was evidently confident of their coherence, but with the hindsight of the position he would come to formulate a little later, their dithering becomes painfully apparent. Algebra for Whewell cannot be *merely* symbolical. Algebraic expressions cannot be empty "equivalent forms" as Peacock portrayed them. To function as a pure Whewellian science, even as a pure meta-science, algebra's symbols must be shown to somehow convey independent content. Like Peacock, he takes that algebra's rules of operation are 'suggested' by lower-level "particular cases". However, he goes on to insist, contrary to Peacock, that once obtained and symbolically articulated, algebra's rules of operation then "define the *meaning* of the symbols," thus lending them—the symbols!—"a generality of signification far wider than existed in the cases which were first represented." Yet at the same time, seeming to realize the hollowness of his claim, reverses the arrow of interpretation, quickly adding that when applied, it is algebra that is interpreted by its lower-level *applicanda* rather than the other way around. When applied, algebraical formulae are matched up to lower-level equivalent forms, but without investing them with any new, independent algebraical content. Content flows in such applications, but exclusively in the other direction, which sits very oddly indeed with his developing doctrine of Fundamental Ideas.

But Whewell says none of this, and continues for another brief paragraph, in which Peacock is referred to directly, as if all is well. "We may easily shew," he goes on to say,

> that irrational, negative and impossible quantities must necessarily be admitted in to our reasonings; and the occurrence of these in the course of our calculation by no means affects the mode of interpreting our results when such quantities have disappeared in our expressions.
>
> Mr. Peacock who has shewn in a very admirable manner the application of the principle above-stated to the various parts of Elementary Algebra, has called it "The Principle of the Permanence of Equivalent Forms"; and expresses it in the following terms.[51]

As if deaf to his own description, Whewell continues to imply that sufficient content flows down from symbolical algebra for irrational, negative, and "impossible" quantities (*sic*) to "be admitted into our reasoning." Just like Peacock, he seems willing to allow symbolical algebra to enrich arithmetical algebra care of the Direct Proposition. But he can only do so at the price of ignoring what for him should be the main question, namely, that of the nature of the *truth* of algebraic expressions in which negatives and imaginaries *are* employed. Such expressions are applied frequently in mathematical physics. For someone who had come by then to vehemently oppose the employment of virtual velocities in mechanics for their failure to correspond to real magnitudes,[52] the question should have been crucial.

Nowhere in the manuscript is the question raised explicitly, but its presence is clearly felt. And in what has to be the manuscript's most revealing passage, Whewell simply rears back. Having evoked Peacock's principle, he first quotes the top-down Direct Proposition:

> "Whatever form is algebraically equivalent to another when expressed in general symbols, must be true whatever these symbols denote" [to which Whewell adds:]
>
> That is when they denote quantities not included in the case in which the equivalence was at first supposed.

However, the Direct Proposition is then *crossed out* and is substituted by a quotation of the bottom-up Converse Proposition!

> "If we discover an equivalent form in Arithmetical Algebra or in any other subordinate science, when the symbols are general in their form though specific in their nature, the same must be an equivalent form when the symbols are general in their nature as much as their form."[53]

Whewell seems to have realized that if the Fundamental Idea ascribed to algebra is taken merely to be that of "absolute generality," algebraic equivalent forms cannot of themselves be considered true in the sense of the term he was developing (namely, like Hamilton, as truth by correspondence to the intuited features of an idea or concept)[54] and are, therefore, powerless to impart any independent content when applied down, rendering the deductive top-to-bottom component of Peacock's principle certainly nonampliative, if not vacuous. The "Converse Proposition" alone remains intact, implying that whatever truth and meaning resides in algebra is funneled up, as it were, from arithmetic. But if that is the case, algebra can no longer be counted a Whewellian pure science. And yet, given its indispensable role, especially in physics, it cannot be simply written out of Whewell's scheme. This is the problem Whewell ran up against in December 1833, from which his philosophy, profoundly as it would be developed, never recovered.

Whewell seems at last to have awoken to the problematic implications of Peacock's position. He refrained from explicitly stating his misgivings in writing, but it is hard to imagine that he and Peacock had not discussed them at Trinity. Nor did Peacock own up to them explicitly, even though in the second edition of the *Treatise*, published during the early 1840s, he too, as we shall see immediately, chose to withdraw the Direct Proposition. Finding himself on extremely precarious grounds, yet unwilling to admit it, Whewell ends the 1833 draft account of the foundations of algebra with the following limp summary:

> The principles which belong to algebra in addition to the more general principle [i.e., that of absolute generality], *are those rules of calculation which it has in common with arithmetic: or rather the rules of arithmetic suggest the rules of combination and corporation of the symbols; while the above principle, imposing the rule of perfect generality on the forms so suggested, is the only imperative and fundamental principle in the science.*[55]

Whewell's induction project for the sciences was embarked upon in order to elucidate the emergence of mathematical physics as the outcome of a meaningful and contrived process of investigation. To this end he had had to dispense with Lagrange's version of the calculus, so fashionable among his peers, precisely because it lacked in extensionality to the extent of rendering its applications justifiable only in retrospect. He had boldly reverted to the discredited limit-based approach, and Kant's equally discredited First Critique, only to discover that algebra itself suffered incurably from the very same malady. By 1840 Whewell had fully realized the futility of integrating Peacock's algebra into his own, by then fully developed, system. Book 2 of the *Philosophy*, entitled "The Philosophy of the Pure Sciences," recapitulates in detail the sections of the 1833 draft that dealt with geometry and arithmetic. Of the meager five pages it devotes to the "Foundations of the Higher Mathematics" (chap. 11), three are dedicated to the idea of limit and only a little more than one to "The Use of General Symbols." There is no real discussion of the nature of algebra, and no attempt is made to apply algebra back down to limits or to motion as in the initial draft. Significantly, a new and essentially historical approach to the question of the truth of arithmetically uninterpretable algebraic formulas makes a brief appearance. The process of interpreting well-formed algebraical results is progressive. Just as the formerly "unmeaning" negative quantities were found "not less significative than positive ones" when applied to Descartes's system of coordinates, so

> as the researches of mathematicians proceeded, other cases also were found, in which the symbols, although destitute of meaning according to the original conventions of their institution, still pointed out truths which could be verified in other ways; as in the cases in which what are called impossible quantities occur.[56]

The direction of "signification" thus remains exclusively bottom up.

As noted above, this would form the cornerstone to DeMorgan's mature position. For Whewell, however, it was not enough to render Peacock's algebra an admissible pure science. Algebra was problematic for Whewell not for being only partly applicable, but because it lacked an *internal* truth relation. No amount of progressive application could change this fact. To this end Hamilton's algebra, founded on a far richer concept of "pure time" than Whewell's arithmetic, would appear to have fared better with Whewell's approach than did Peacock's (or DeMorgan's, for that matter). It is highly improbable that Whewell was unacquainted with Hamilton's algebraic work, and yet nowhere in his writings is it even mentioned. Hamilton, as Hankins observes, remained "in splendid isolation."[57] Whewell either found something very wrong with Hamilton's algebra (perhaps the fact that it was noncommutative), or had lost the inclination to develop or modify his philosophy any further by the time Hamilton's work on quaternions was published during the late 1840s. I suspect, but cannot confirm, that the latter was the case.[58]

PEACOCK'S SELF-CRITIQUE

By 1842, Peacock had apparently also arrived at the same conclusion. In the preface to the first volume of the second edition of the *Treatise*, the Direct Proposition is dropped entirely and the "Principle of the Permanence of Equivalent Forms" is expressed as follows:

> It is this adoption of the rules of the operations of arithmetical algebra as the rules for performing the operations which bear the same names in symbolical algebra, which secures the absolute identity of the results in the two sciences as far as they exist in common: or in other words, all the results of arithmetical algebra which are deduced by the application of its rules, and which are general in form, though particular in value, are results likewise of symbolical algebra, where they are general in value as well as in form.[59]

As I believe was the case in Whewell's draft of 1833, the elimination of the Direct Proposition in the second edition of the *Treatise* is much more than a technicality, and it has a profound effect on the entire structure of the work, which, I believe, signals a significant change of heart on Peacock's behalf.

The second edition was produced in two volumes (published three years apart) in which arithmetical and symbolical algebra are treated separately. In them the "Principle of the Permanence of Equivalent Forms" is discussed twice: once in the preface to volume 1,[60] and again in a short chapter devoted to its "formal statement" in volume 2.[61] In both cases it comprises one proposition only, the one referred to as "Converse" in both the first edition and the BAAS Report of 1834. It is described in the preface to volume 1 as the general principle that expresses "the general law of transition from the results of arithmetical to those of symbolical algebra" with no mention made of subsequently traversing the divide in the opposite direction. "Upon this view of the principles of symbolical algebra," he continues, no longer mincing his words, "it will follow that its operations are determined by the definitions of arithmetical algebra, as far as they proceed in common."[62] Nowhere throughout the two volumes is the question of applying the results of symbolical algebra back down to arithmetical algebra even raised.

Volume 2 speaks much of applying new results of symbolical algebra—not back to arithmetical algebra, however, but to geometry and trigonometry (or as Peacock prefers: "goniometry"). The overall effect is quite impressive, and a major step away from the *Treatise* of fifteen years earlier. Arithmetical algebra is still made to "suggest" symbolical algebra as before, but the differences between the two editions are profound. The book no longer reasons back and forth across the great divide between the two algebras in the bootstrapping manner of the first edition. Arithmetical algebra is first fully developed as an autonomous, self-contained system. Only then, in a separate volume, and in a manner no longer related to number, is it further generalized, restated formally, and declared the basis for a fully fledged symbolical system, which from that point onward is developed in its own right, from its own principles, and whose quality, strength, and value are measured by its formal completeness, on the one hand, and the richness of its application, on the other.

Symbolical algebra is no longer presented as a question-begging operation for the sake of enriching arithmetical algebra, but as a formal, logically connected system, lacking intrinsic meaning or a truth relation, for which the science of number (arithmetical algebra) is considered but one of its several (partial) models—and that of geometry the most important. With symbolical algebra fully developed and in place, and its areas capable of arithmetical interpretation (those initially "suggested" by arithmetical algebra) well charted, the idea is then to cast around for other models capable of rendering other portions meaningful. Unlike the *Treatise* of 1830, the second edition no longer regards the science of number as algebra's main beneficiary in this respect. Rather than those of number, "it is from the relations of space," announces Peacock at the outset of volume 2, "that Symbolical Algebra[63] derives its largest range of interpretations, as well as the chief sources of its power of dealing with those branches of science and natural philosophy which are essentially connected with them." Indeed, Peacock's most significant contribution to mathematics is the important anticipation of vector analysis found in the geometrical interpretation of imaginaries dealt with at length in volume 2.[64] Peacock seems well aware of the significance of his move, for he continues:

> it is for this reason that I have endeavoured to associate Algebra with Geometry throughout the whole course of its development, beginning with the geometrical interpretation of the signs + and − when considered with reference to each other, and advancing to that of the various other signs which are symbolized by the roots of 1: we are thus enabled, in the present volume, to bring the Geometry of Position, embracing the whole theory of lines considered both in relative position and magnitude and the properties of rectilinear figures, under the Dominion of Algebra.[65]

Of course, a prudent "Pauline" theoretician would find little point in developing symbolical algebra to this end. For him the value and worth of a field of mathematics is measured by the conceptual clarity of its foundations and the logical connectivity of what follows from them. Hamilton's quaternions project is precisely an attempt to construct a fully fledged algebra without having first to construct a purely formal, conceptually vacuous system to serve as its "dominion." But to this end Hamilton was willing to do what Peacock, I believe, never considered: namely, to modify and enrich the very notion of number. In 1842 Peacock fully admits that the original *Treatise* of 1830 had been seriously flawed, and that, despite their shared title, the new work is not a second edition in the simple sense of the term.

> It is true that the same views of the relations of the principles of Arithmetical and Symbolical Algebra formed the basis of my first publication on Algebra in 1830: but not only was the nature of the dependence of Symbolical upon Arithmetical Algebra very imperfectly developed in that work, but no sufficient attempt was made to reduce its principles and their applications to a complete and regular system, all whose parts were connected with each other: they have consequently been sometimes controverted upon grounds more or less erroneous.[66]

By 1840 or so, Peacock had clearly made his peace with the idea of an autonomous and genuinely symbolical algebra, which, though initially suggested by a system of generalized arithmetic along the lines of Frend and Maseres, provided syntactical grounding for a variety of mathematical applications much in the spirit of Babbage's initial suggestions. The dependence of Peacock's algebra on arithmetic remained problematic in ways that Babbage's system did not, but his talk of its autonomy was no longer mere lip-service, and the formative role he gave it with respect to geometry was real.

The problem was that by 1845 Peacock's system, improved as it may have been, was no longer news. With the development of abstract algebra well under way in the good hands of DeMorgan, Gregory, and later Boole, and with Hamilton's quaternions project approaching completion (and resembling less and less the meaningful, theoretical enterprise it had set out to become), Peacock's new and improved textbook no longer represented the cutting edge of meta-mathematical deliberation. It broke no new ground. Peacock, one could say, reacted to his own system too slowly for his second thoughts to be much relevant to the different ways in which his colleagues and contemporaries had reacted to it. And thus, the work for which he is justly remembered is the gravely flawed version of 1830, rather than the better thought-out, mathematically more viable refashioning of it fifteen years later.

This is not to be regretted, however. Flawed as it was, Peacock's original *Treatise on Algebra*, as I have argued, was a bold, pained, creative attempt to negotiate a dilemma he was incapable of resolving. Unable to decide the issue this way or that, the solution he proposed in effect preserved the impasse rather than breaking it. For Peacock, and those who shared his Lockean approach to mathematics, intellectual integrity, and deep knowledge and appreciation of continental analysis, traditional portrayals of algebra as generalized arithmetic were as unacceptable as the type of radical formal operationalism that Babbage had proposed. The novelty and significance of the *Treatise* becomes apparent when one realizes, as this chapter has argued, the extent to which it consciously aspired to forge a creative compromise between these two alternatives, of which its author, we now know, was all too well aware. The compromise itself was exceedingly problematic. It never stood a chance. The importance of the *Treatise* to the history of mathematics lies, then, not in the particular dualism it offered, as much as in the way in which the dualism it proposed preserved, drew attention to, and further emphasized the very dilemma it was designed to resolve! Peacock had no disciples, and the tale of two algebras he proposed was taken up by no one, except briefly by Whewell. To those unaware of Babbage's unpublished work, Peacock's symbolical algebra might have been their first exposure to algebraic formalism (DeMorgan may have been thus influenced). But for the larger part, the *Treatise*'s real impact lay in the way it so forcefully raised the problem, dithering undecidedly, at is did, between the two seemingly unacceptable and irreconcilable normative alternatives it purported to straddle.

The chapters comprising part 2 of this book argued for a way in which a scientific framework transition could be viably considered rational. It envisaged a two-stage process in which a practitioner of voice and standing was first rendered ambivalent toward the framework constitutive of his thinking by exposure to external normative criticism, and then succeeded unwittingly to further ambivalate key players within his community by means of his attempt creatively to overcome his predicament. The story related in the three chapters that make up part 3 of this study, on the making and impact of Peacock's *Treatise* of 1830, sought to exemplify such a story in some detail. Chapter 5 first analyzed the specific challenges to his initial meta-mathematical Lockean commitments he encountered in the "trading zone" of the Analytical Society, culminating in the stark formalist alternative presented to him by Babbage. Chapter 6 presented close readings of three texts penned by Peacock in response: his notes to Lacroix, his arithmetic project, and finally the *Treatise* itself. The notes were read as motivating his search for an account of algebra that would do justice to the fullness of the development of the calculus without transcending the confines of his Lockean commitments. The arithmetic project was read as a heroic failed attempt to do so. The *Treatise* was then read as a brave and highly creative attempt to reach beyond those confines by fashioning a theory of algebra that comprised both generalized arithmetic and symbolical components in a tense and hardly coherent symbiosis, as a classical example of the type of unstable hybridic work capable of ambivalating other key players. Chapter 7 then went on to briefly chart the way in which DeMorgan, Hamilton, Gregory, and Whewell each read the book, and responded to it. The new views of algebra they formulated in response, with the exception of Whewell, were shown to go important steps toward replacing the profound ambivalence that plagued the *Treatise* with exceedingly more coherent accounts of the foundations of algebra that veered clear of the *Treatise*'s Lockean moorings toward a less convoluted endorsement of a formalist approach. Peacock was then shown to have followed suit, though too late to make a difference.

The story of this fascinating moment in the history of mathematics has been told over the years by the several historians of mathematics referred to and engaged with in the text. But it has never been told as it is told here, as a paradigmatic example of the crucial role played in major framework transitions by the creative indecision bred of the profound ambivalence of an exemplary scholar and thinker.

ACKNOWLEDGMENTS

I began writing *Creatively Undecided* in earnest upon arriving in Princeton in the fall of 2010 as a member of the Institute for Advanced Study's School of Historical Studies. By the time I left Princeton for Collegium Budapest in April 2011, I had completed first full drafts of the first four chapters, and during my five months in Budapest (which ended with the original Collegium's much regretted demise in August 2011), I completed much of the initial writing of part 3. In their different ways both institutions offered ideal environments and working conditions for such an undertaking. It is very difficult to find the time and freedom of mind required for thinking through such a project and committing it to writing in the normal course even of our privileged employments as university professors—all the more so in the distressingly trying circumstances of the country from which I come. My profound gratitude for the opportunity to spend a full twelve-month sabbatical between the IAS and the late Collegium Budapest cannot be adequately expressed in words. This book owes to them decisively.

During the four years between returning from Budapest and submitting the work to University of Chicago Press in August 2015, I returned to the manuscript sporadically, but allowed myself to be distracted by other projects. This was mainly due to being deterred by the enormity of what I had committed myself to in part 3. I knew by then what it took to tell the story of Peacock's algebra the way I wanted to, but I could not see a manageable way around doing the same for the other three stories I had planned to tell. The penny dropped sometime during the early spring of 2015 in clear realization that my original plan for part 3 merited a separate book; that it was impossible to do justice to the philosophical undertaking of parts 1 and 2 and to the fourfold story of Peacock's, Herschel's, Whewell's, and Hamilton's hybridic meta-scientific musings within the confines of a single work.

What was needed for part 3, I realized with relief, was a substantial case study for which the Peacock story should be more than enough. Setting it alongside the other three, and weaving them into a single tapestry, would have to wait its turn.

Creatively Undecided owes much to many. As noted in chapter 1, I had awoken to the strangely split and hybridic writings of my four early Victorians almost a decade and a half before embarking on this book, and I began thinking seriously about their philosophical significance a few years later. Early papers related to the former were presented at the Dibner Institute, MIT Fellows Colloquium, the Bar Hillel Colloquium, Jerusalem, and HPS colloquia at Bar Ilan University and the University of Notre Dame. Papers on elements of the latter were delivered as they emerged from 2004 onward, again at Notre Dame (as part of the Crown-Minow Lecture Series I delivered there in 2004), the Graduate Program for Interpretation and Cultural Studies and the Philosophy Department colloquia at Bar Ilan University, again at the Bar Hillel Colloquium, and at HPS seminars at Princeton and Frankfurt, at conferences at the Van Leer Institute and the Hebrew University, Jerusalem, and at the universities of Ghent, Bergamo, and Tel Aviv, as well as at guest lectures at the departments of biomedical engineering at Tel Aviv and biochemistry at the Weizmann Institute. I thank my hosts and audiences in all these venues for the challenge of having to articulate my thoughts out loud to them, and to respond to their many comments, queries, and criticisms. Looking back, it feels as if the entire project took proper form less in actually writing it up than in the course of talking about it with an extraordinarily bright and varied spectrum of colleagues, students, and friends.

Several individuals read and responded to portions of this study at various stages of its coming together. Special thanks go in this regard to Ori Belkind, Yitzhak Benbaji, William Berkson, Daniel R. Brooks, Leo Corry, Lorraine Daston, Yuval Dolev, Noah J. Efron, Paul Franks, Eli Friedlander, Michael Friedman, Ariel Furstenberg, Daniel Garber, Snait Gissis, Jeremy Gray, Niccolò Guicciardini, Zoe Gutzheit, Oren Harman, Don Howard, Kevin Lambert, Ehud Lamm, Anat Matar, Elijah Millgram, Joan L. Richards, Sheldon Richmond, Sam S. Schweber, Orly R. Shenkar, Alfred I. Tauber, Hami Verbin, Jeremy Wanderer, and Ido Yavetz; and to my students Yael Gazit, Assaf Heller, Omer Michaelis, Yaara Rosolio, and Yonatan Shaham.

Last but not least, I would like to warmly thank Karen Merikangas Darling, senior acquisitions editor for science studies at the University of Chicago Press, for her expert handling of the manuscript, and the two anonymous referees she enlisted, whose thorough and detailed readings (and occasional misreadings), comments, queries, and suggestions enabled me to reread my text through the eyes of its ideal readers: richly informed and philosophically motivated scholars interested in science, appreciative, yet interestingly wary of what I was trying to do. Special thanks also goes to Lys Ann Weiss for her superb copyediting to which the book's final version owes enormously. Thank you, Karen, thank you, dear readers, and thank you, Lys!

This book is dedicated with love to the Cohn Institute for the History and Philosophy of Science and Ideas at Tel Aviv, which has been my academic home since granting my Ph.D. (the Cohn Institute's first) over thirty years ago—to its faculty, staff, students, and the blessed memory of its two cofounders. Thanks to the great generosity of Bert and Barbara Cohn, the Cohn Institute has succeeded over the years in manifesting and extending the late Yehuda Elkana and Amos Funkenstein's bold vision for the study of science in the richness of its diverse contexts and as refracted through the lens of the entire range of the humanities and social sciences. In this age of narrowing specialization and immediate impact factoring, the kind of freedom, so valued and encouraged by the Cohn Institute, to ponder far afield for years in order to rethink the basics of one's discipline rather than eagerly apply them, should not be taken for granted. I doubt I would have been motivated in the same way to undertake elsewhere this work, whose conclusions in a deep sense refute current trends of academic evaluation.

NOTES

INTRODUCTION

1. On the attempts by Agassi, Lakatos, and Bartley to address it from a more Popperian perspective, see below, chapter 1, note 37, and chapter 2, notes 10 and 11 and accompanying text.

2. The allusion to Habermas is deliberate. Friedman constructs his account on the basis of Habermas's notion of communicative rationality.

3. Comparative Irrealism—the idea that, for lack of a shared vocabulary of thin normative concepts (such as "good", "wrong" and "reasonable"), sufficiently diverse normative frameworks can never be compared in a manner agreeable to both—is argued for in Fisch and Benbaji 2011, chaps. 2 and 3. See also chap. 5 of that work for a detailed critique of Friedman's account.

4. I am thinking mainly of Korsgaard 1996a and 2009; Frankfurt 1988, 2004, and 2006; and the papers on human agency and personhood collected in Taylor 1985a and 1985b, culminating in part 1 of Taylor 1989.

5. See in particular Brandom 1994, 2000a, and 2002, and Walzer 1987.

6. The set of three lectures is entitled "Animating Ideas of Idealism: A Semantic Sonata in Kant and Hegel," and was published in Brandom 2009, chaps. 1–3.

7. Brandom 2009, 36–37.

8. Brandom 2009, 62.

9. Brandom 2009, 68.

10. One possible reason for Brandom's oversight is his insistence on limiting his discussion of normativity to *conceptual* norms, and, not unlike McDowell, on articulating meaning and intentionality, rather than truth. It is true that human sapience owes almost exclusively to our conceptual capacities, but focusing on language alone blinkers his discussion from the phenomenon and challenges of normative diversity. Among speakers who share a language, there may be vast differences of opinion and commitment, but very little diversity of conceptual norms per se.

11. Precisely how this works is argued in detail on the basis of some of Frankfurt's work on agency in chapter 8 of Fisch and Benbaji 2011, and briefly reiterated in Fisch 2013b and in the latter part of chapter 3, below.

12. I am referring in particular to Fisch 1985, 1987, 1988, and 1991a. For brief accounts of how I stand with regard to the work done on Whewell's philosophy since then, see below, chapter 1, note 46, and chapter 5, note 25 and accompanying text, for Laura Snyder's account of his philosophy; and chapter 1, note 45, and chapter 7, note 36 and accompanying text, for that of Stephen Ducheyne.

CHAPTER ONE

1. Fisch 1999. Whewell, as we shall see more fully below, seems to have briefly endorsed Peacock's view of algebra as the basis for a treatise of his own, "The Philosophy of Pure Sciences" (Cambridge, Trinity College Library, Whewell Papers (hereafter, WP), R.18.178). But by late 1833 or so the project was aborted (the last entry is dated December 6, 1833) in favor of a text by the same name that would become book 2 of his *Philosophy of the Inductive Sciences* (Whewell 1847, 82–163), in which questions concerning the foundations of algebra that were raised following Peacock in the manuscript version of 1833 are elegantly sidestepped. For an initial analysis of the manuscript and the possible reasons for its discard, see Fisch 1994c, §5.

2. See especially Fisch 1985, 1987, 1988, 1991a, 1991b, 1991c, and 1994c.

3. Fisch 1992, 1994a, 1994b, and 1997, chap.1.

4. On the extent to which "scientific and technological knowledges . . . and the means by which they are produced are widely accorded impersonal" and the need to redress the balance by showing "how and why people and their virtues matter to [their] making and the authority," if from a different perspective, see Shapin 2008, especially chap.1.

5. Lakatos 1971.

6. Lakatos acknowledges four such methodologies: Baconian inductivism, instrumentalism, Popperian falsificationism, and his very own "methodology of scientific research programmes." For criticisms of Lakatos's approach, see Curtis 1986, Garber 1986, and Fisch 1994a and 1994b. For discussion of the Hegelian aspects of Lakatos's historiography (the way rationality is conferred upon one's current position by rationally reconstructing the tradition to which it belongs), see Hacking 1979/1981. For a comparison of Lakatos's to Brandom's (2002 and 2009) brand of rational reconstruction, see Fisch and Benbaji 2011, chap.7, §1. One major difference between the two accounts is that for Lakatos it is ultimately the historian's latter-day philosophical position that confers rationality on the past, while for Brandom, who steers far closer to Hegel in this respect, it is the decisions of the past that confer rationality on the normative rulings of the present.

7. Collingwood 1939, chap. 5; Fisch 1991a, chap. 1.

8. In the following oft-quoted passage, Collingwood compares trying to read Plato's *Parmenides* to trying to understand the structure and functions of a Roman fort. In both cases, all one has to work with is a text or artifact whose meaning is wholly determined by the erotetic process of self-questioning that originally motivated its particular formation:

"'Sblood' says Hamlet 'do you think I am easier to be played on than a pipe?' Those eminent philosophers, Rosencrantz and Guildenstern, think tout bonnement that they can discover what the Parmenides is about by merely reading it; but if you took them to the south gate of Housesteads and said, 'Please distinguish the various periods of construction here, and explain what purpose the builders of each period had in mind,' they would protest 'Believe me, I cannot.' Do they think the Parmenides is easier to understand than a rotten little Roman fort? 'Sblood!'" (1939, 40).

9. Hamilton to Whewell, 25 May 1833, WP Add.MS.a.205[102]. Cited by Graves 1885, 48–49, and Hankins 1980, 175; analyzed by Fisch 1991a, 64–67.

10. Whewell 1844 and 1847, 1: chap. 2.

11. With the possible exception of Charles Sanders Peirce, whose 1869 unpublished Harvard "Lectures on British Logicians" clearly register both true understanding and enthusiastic endorsement of Whewell's "inductive" approach to philosophy of science. All mention of Whewell in this respect vanishes, however, in Peirce's published work. For further details, and a possible explanation of Peirce's reasons for concealing his debt to Whewell, see Fisch 1991a, 106–10.

12. For an initial outline of the historiographical issues at play, see Fisch 1994c, especially §2.

13. Korsgaard 1996a, 93.

14. The idea permeates the writings of John McDowell and Robert Brandom, at least since the publication in 1994 of the former's *Mind and World* and the latter's *Making It Explicit*. Friedman 2001 refers to the term explicitly only once (85). However, the distinctly communal setting of Friedman's discussion of scientific rationality in the light of Kuhn, and especially the resource it finds in Habermas's notion of communicative rationality, renders the term central to the book's entire project. In Friedman's later work the term is employed more frequently, but loses its original discursive connotation by morphing interestingly and consistently from talk of a "space of [intellectual or empirical] *reasons*" to that of "[intellectual or empirical] *possibilities*," thus marking, as I shall argue below, a retreat from his original focus on questions of agency and rationality. For the new phrasing see, for example, Friedman 2010a, 714, 717, 721.

15. "In characterizing an episode or a state as that of *knowing*, we are not giving an empirical description of that episode or state; we are placing it in the logical space of reasons, of justifying and being able to justify what one says" (Sellars 1956, §36).

16. As Brandom 2002, 6, puts it: "to *be* rational *is* to play the game of giving and asking for reasons" (italics added).

17. Habermas 2000, 323.

18. Habermas 2000, 323.

19. For a recent reading of Hegel along these lines, see Pippin 2011, 37–39 and *passim*.

20. Brandom 2002, chap. 1. The five are the logical, the instrumental, the translational (or interpretive in Davidson's sense of the term), the inferential, and the historicist conceptions or views of rationality.

21. Brandom 2009, 2–3 (emphasis added to all but *reasons*).

22. Frankfurt 1987, 172–73.

23. Korsgaard 2009.

24. I myself belong to the school of thought that deems *all* facts framework dependent. Popper obviously does not. "In defense of science and rationality," as his subtitle reads, Popper (1976 and 1994) famously condemns the idea of Framework Dependency as

a "myth." For the notion of framework-dependent factuality, even in ethics, see Fisch and Benbaji 2011, especially chaps. 2–4.

25. Criticism, I shall argue, involves more than merely *exposing* problems. A negative test result does not amount to criticism unless *someone* is made responsible for the failing. A system can be tested and found problematic, but only another agent can be criticized. See chapter 3, below.

26. Korsgaard 1996a, 68.

27. Fisch and Benbaji 2011, chap. 6, §1.

28. McDowell 1994, 81.

29. McDowell 1995, 101–2.

30. See, in this regard, Charles Taylor's idea of the role of "hypergoods" as normatively governing lower-level commitments (Taylor 1989, chaps. 3 and 4), and its criticism by Anderson (1996, n. 24).

31. See, for instance, Kuhn 1993, 338–39.

32. See, for example, Kuhn's reply to McMullin's claim that he, Kuhn, denies the "epistemic character" of scientific theories.

"My goal is double. On the one hand, I aim to justify claims that science is cognitive, that its product is knowledge of nature, and that the criteria it uses in evaluating beliefs are in that sense epistemic. But on the other, I aim to deny all meaning to claims that successive scientific beliefs become more and more probable or better and better approximations to the truth and simultaneously to suggest that the subject of truth claims cannot be a relation between beliefs and a putatively mind-independent or 'external' world" (Kuhn 1993, 330).

33. Popper's criticism of instrumentalism permeates his early work. For one pertinent discussion, see the second of the "Three Views concerning Human Knowledge" (Popper 1963, chap. 3).

34. Agassi 1964 stands out among the work of Popperians in early acknowledging "the root" of scientific problems in metaphysics, and extending Popperian fallibilism to the metaphysical level. However, this otherwise important paper fails to acknowledge the constitutive normative hold the kind of worldviews he discusses have on the experience and thinking of their adherents. He writes as follows:

> Two metaphysical views offer two different interpretations of a body of fact. Each of these interpretations is developed into a scientific theory, and one of the two scientific theories is defeated by a crucial experiment. The metaphysics behind the defeated scientific theory loses its interpretative power and is then abandoned. (191–92)

The logic is thoroughly instrumental: Metaphysical systems offer interpretations of bodies of fact, and can be developed into scientific theories about those facts, which, if and when empirically refuted, fatally fault the metaphysical system that begot them. The argument assumes a shared recognizable distinction between the body of fact in question and its various metaphysical interpretations, as if truly committed Aristotelians and Newtonians experience the world alike, as if part of the conceptual content that informs experience can be considered neutral across metaphysical commitment in a manner relevant to its refutation. (For a more detailed argument against this type of perspectivism with regard to normativity in general, see Benbaji and Fisch 2004, §1.2; and Fisch and Benbaji 2011, chap. 2, §2.)

In addition to underestimating a worldview's constitutive force regarding the factual (see note 24, above), Agassi underestimates, in my opinion, the normative hold a truly internalized metaphysical worldview has on those whose worldview it is. (With the possible exception of a brief remark on the need for cooperation in science, because a person [Robinson Crusoe] working alone could not advance much "because of his limited capacity to criticize himself and thus get out of the routine of his way of thinking" (190).) Such commitments are not easily abandoned—not because of habit, or laziness, or lapses in rationality, but because in becoming second nature (to use McDowell's term), they *commit* one to the interpretation of the world and self to which they give rise. Combining normative hold and constitutive force, metaphysical worldviews of the kind Agassi discusses are definitive of the very reasoning he maintains can be used against them. For a clear and appreciative exposition of Agassi's position, see Berkson 1995.

35. As Hamilton would describe the first of the two in a paper following Kant, entitled "Algebra as the Science of Pure Time" (Hamilton 1837, 1–6), that heralded what would become Hamilton's quaternions project. Though published in 1837, the paper was read in two installments at the Royal Irish Academy as early as 4 November 1833 and 1 June 1835. In a letter to Peacock dated 23 July 1835 (Cambridge, Trinity College Library, Add.ms.b.4944), Hamilton describes his "nearly printed essay" as being written in direct reaction to reading Peacock's *Treatise*. We shall return to this paper and its relation to Peacock in chapter 6.

36. "Essays on the Philosophy of Analysis," London, British Museum, Add. MS 37202.

37. Peacock to Babbage, 7 May 1822, London, British Museum, Add. MS 37182, No. 323, cited in Dubbey 1978, 95.

38. See, for instance, Kuhn 1970, 94, 150, 151.

39. *Pace* Agassi 1964; see note 34, above.

40. See especially Herschel 1832. Apparently his commitment to Lagrange dated back to his schooldays. As a teenager, Herschel was tutored privately in mathematics to prepare him for university by a Scottish teacher named Rogers (Buttmann 1970, 9; Snyder 2011, 28). Ashworth (1996, 636) identifies him as Edinburgh-trained accountant and mathematician Alexander Rogers, whose letters of 1808 and 1809 to the sixteen-year-old Herschel recommended his reading Lacroix's French textbook on the differential and integral calculus, as well as Lagrange's *Mécanique analytique*.

41. For Herschel's early attraction to Bacon's philosophy, see Snyder 2011, 38–43, and references there. For his mature position, see especially his thoroughly and explicitly Baconian *Preliminary Discourse on the Study of Natural Philosophy* (Herschel 1830).

42. The tension evident in Herschel's works between his Baconian view of science and Lagrangian view of mathematics has little to do with the renewed dispute among historians regarding Bacon's attitude to mathematics in science, and his consequent influence on Newton. The standard view, introduced by Koyré 1939 and Kuhn 1977 and voiced more recently by Gaukroger 2001, 25–27, portrays Bacon as wholly dismissive of mathematization in science, and hence of neglible influence on the emergence of Newtonian physics. The standard view has been convincingly called into question, first in Rees 1985

and 1986 and more recently by Domski 2013. The problem Herschel grappled with and despaired of solving, however, was that of squaring his Baconianism with a *thoroughly formalistic* view of mathematics, which was never a seventeenth-century option.

43. See Hankins 1980, chap. 27; and especially Yeo 1993, 67–71; and Brown 2012 and 2013, 1–6, and references there.

44. To which he added Faraday's electrochemistry in later years.

45. An ill-worded section in my early book on Whewell (Fisch 1991, §3.1221) created the false impression, rightly called to task by Ducheyne (2011 and 2014), that Whewell's philosophy of science was unrelated to the concerns and content of Kant's First Critique. Given the ample references to Kant in that work, nothing could be further from the truth. What I had meant to argue in that section (entitled "Kantian Influence Set Aside"), was *not* that Whewell's mature philosophy did not bear profound affinities to Kant's, or that he had not read Kant extensively, but rather, as I put it in a lengthy footnote, "the fact that his view of science approximated to that of Kant in certain respects does not itself imply that it was reached directly or indirectly through Kantian *influence*" (Fisch 1991, 192n6 [italics added]). Writing from a squarely Popperian perspective, I sought to reconstruct Whewell's shifts and turns en route to his final position as creative attempts to tackle problems he encountered, rather than as the result of influence, even if he ended up with a position that to an extent resembled that of one whom he described as "the great metaphysical genius in Germany" (Whewell 1840, 165).

46. It is evident from even this brief formulation that I remain largely committed to the account of Whewell's philosophy and its development that I published in the early 1990s. Since then the literature on Whewell's philosophy has grown considerably, thanks in particular to the large body of work Laura Snyder has devoted to his thinking in her important studies, first of the philosophical debate between him and Mill (Snyder 2006) and more recently of the evolving friendship and impact of the four members of the "Philosophical Breakfast Club"—Herschel, Babbage, Whewell, and Jones (Snyder 2011). Snyder's understanding of Whewell's philosophy of science, its development and significance, and its relationship to the thought of Bacon and Kant differs from my own, as does Ducheyne's more recent attempt, noted above, to interpret Whewell as grappling with Kant. Neither Snyder nor Ducheyne has taken due notice of the inherent ambivalence of Whewell's mature philosophy or of its possible sources.

47. For a detailed description of my work with Benbaji, and the ambivalating effect on it of another colleague, see Fisch 2015.

48. See note 34, above.

49. Wittgenstein 1974, 80e–81e; e.g., Rorty 1989, 49; Davidson 1982.

50. I am referring especially to Brandom 1994, the account of norm changing outlined and exemplified in Brandom 2002, and developed in Brandom 2009, chaps. 2–3; Walzer 1987 and 1988, and Sreenivasan 1998, the most important development of Walzer's "interpretive" approach; and the Habermasian account of "the distinctive intersubjective rationality" that framework transitions in the exact sciences "can justly claim," first proposed in Friedman 2001 and rearticulated in Friedman 2010b (the quoted sentence is taken from its opening paragraph).

51. I am referring to his early expressivist work on language, agency, and the sciences of man: Taylor 1971 and 1977, as well as Taylor 1985a, chaps. 2, 4, 9 and 10.

52. Especially Korsgaard 1996a, chaps. 2–3, and 2009, chaps. 1–4.

53. All Frankfurt's writing on personal identity and the will since his seminal works (Frankfurt 1971 and 1982) are relevant. For a concise and well-rounded rearticulation of his position, see Frankfurt 2004 and 2006.

54. Which is as close as we get to Walzer's notion of "connected" criticism.

55. Wittgenstein 1974, §609–12.

56. The chapter in question is partly anticipated in Fisch 2008, and somewhat enlarged on in Fisch 2010.

CHAPTER TWO

1. In response to the worry of one of the two anonymous readers of the book for the press.

2. Rorty (1989) adopts Bloom's (1973) depiction of "strong poets" as centrally motivated by the anxiety of unoriginality, as the way to account for paradigm shifts in general.

3. Hence, as mentioned, the stress that some of the thinkers discussed below place on intentionality rather than truth.

4. Like that of a normative outlook, the constitutive force of scientific frameworks extends beyond the bodies of scientific knowledge proper that their practitioners produce, to the dispositions, skills, techniques, and practices *by which* they are produced. In what follows, I shall use "scientific knowledge" generously to include both knowing-that and knowing-how.

5. Quine 1990, 15.

6. For Friedman's effective criticism of Quine 1951, see Friedman 2001, 32–34.

7. For the interesting halfway position between Quine and Friedman fashioned by Putman, also in response to Quine 1951, see Putnam 1976; Tsou 2010; and chapter 3, note 8, below.

8. Friedman 2001, 35.

9. Agassi 1964. See chapter 1, note 37, above.

10. A similar flaw plagues the discussion of commitment in Bartley 1984, which at first blush would seem to be animated by the concerns of the present essay far more than Agassi or Lakatos. Bartley casts his net much wider than Agassi or Lakatos in an attempt quite generally to refute the idea that in all human endeavor, including religion, ethics, politics, economics, and science, "there exists an essential logical limitation to rationality; that rational defense and examination of ideas must, for logical reasons, be terminated by an arbitrary and irrational appeal to . . . dogmas or absolute presuppositions" that are "(1) chosen arbitrarily and irrationally, or forced on one by the circumstances of fate or history, marked the limits of rationality"; are (2) "not subject to review or criticism"; and are (3) "incapable of justification." And "(4) all of one's positions that were rational were justified or defended in terms of these presuppositions or dogmas: that is, any rationality in one's life was rationality relative to irrational bases" (Bartley 1984, app. 3). I of course sympathize with Bartley's identification of rationality and criticism, as well as with his philosophical apprehension (though not with his condescending description of rationality's precommitments). I part ways with him decisively, however, on two

major points that will become more apparent as we proceed: first, in insisting that criticism, in purporting to expose normative failings (and not merely address inconsistencies), *necessarily* presupposes commitment to a normative system; and second, that breaking the hold of such commitment rationally cannot be resolved by what Bartley dubs "competition" (see the following note). I find his notion of criticism surprisingly bland in comparison to the liveliness of his discussion, and his viewing of serious normative commitment as something one can adopt, reject, or retreat to at will I find unserious to the point of irrelevancy.

11. Chapters 2 and 3 of Fisch and Benbaji 2011 are devoted to making the case for Comparative Irrealism—the claim that for lack of a shared vocabulary of thin normative concepts, sufficiently diverse normative outlooks are normatively incomparable in a manner acceptable to both. The argument is essentially semantic and leveled both from communal and idiolect perspectives on meaning, respectively.

12. To which a third, intersubjective self-critical element was added, pertaining to their discussion with others—an element to which we shall return in connection with the third, Hegelian bias that grounds this study.

13. There is, hence, a crucial difference between acting *for* a reason—that is, taking action *because* one deems it right—and justifying it after the event. Only the former counts as acting rationally. The latter, by contrast, is to retrospectively *rationalize* one's otherwise thoughtless, or wrongly motivated, actions—their utility and beneficence notwithstanding. Korsgaard clearly formulates what she calls "the problem of the normative" in terms of this sense of the priority of reason to action. Sellars, Brandom, and McDowell, who are less interested than Korsgaard in agency proper, locate rationality in the essentially timeless inferential *relations* obtaining at any one moment within a person's set of commitments (or as Brandom now calls it following Kant, "synthetic unity of apperception" [2009, 41 and following]), regardless of whether reason prompted action prospectively, or action retrospectively prompted reasoning.

14. Taylor 1985a, 3.

15. "The gradual invention of opera," writes Pippin (2008, 75) convincingly, "cannot be said to be something someone or some group intentionally invented in some singular act, but it *was* invented, made." But accounting for the invention or radical modification of other frameworks, such as classical physics, the formalist conception of algebra, and even that of the Hegelian approach to rational agency to which Pippin's own book is devoted, is not as easily dismissed.

16. Frankfurt 2006, 4.

17. Frankfurt's (1971) now famous distinction between the willing (or wanton) and unwilling addict brings this out forcefully.

18. Korsgaard 1996a, 97–113.

19. On the "un-Kantian" element in Korsgaard's approach, see Onora O'Neill's introduction to Korsgaard 1996a, xiv; and Nagel 1996, 203–4. For a different view, and a critique of how Korsgaard's "practical identities" stand to Kantian morality, see Pippin 2008, 85–90.

20. Korsgaard 1996a, 101, and 2009, 20–25.

21. Korsgaard 1996a, 101.

22. This work has its beginnings in Taylor 1964 but only reaches fruition when the

crucial linguistic element is added, especially in Taylor 1971, 1977, and 1985a, chap. 2. Taylor's intense, career-long engagement of the unique linguistic aspects of the nature of human selfhood culminates majestically in Taylor 2016.

23. Taylor 1977, 19–21.

24. Taylor 1977, 19.

25. See Taylor 1981, 105–7.

26. Taylor 1977, 19.

27. See especially Taylor 1971 and the joint introduction to Taylor 1985a and 1985b.

28. Taylor 2016 is a systematic defense of the romantic or "constitutive" view of language inaugurated by Hamann, Herder, and Humboldt (and dubbed the HHH theory) in the face of an equally systematic critique of what he calls the "enframing" or "designative" theory of language inaugurated by Hobbes, Locke, and Condillac (appropriately dubbed HLC). For one of several powerful early statements of the romantic element in his analysis, see Taylor 1985a, chap. 9.

29. At one point in *Sources of the Self*, Taylor interestingly compares how, in ethics as in science, awareness of the process of "historical supersession of less adequate views," by which the framework within which we stand replaced earlier ones, serves to establish it "as a standard of criticism of contemporary beliefs and practices" (Taylor 1989, 64–65 and 68–69).

30. See especially Taylor 1989, chap. 3.

31. Taylor 1989, 63.

32. Taylor 1989, 71.

33. Taylor 1989, 73.

34. For a friendly yet pointed critique of Taylor's view of science along these very lines, see Geertz 1994.

35. See Benbaji and Fisch 2004 and 2005; and Fisch and Benbaji 2011, chaps. 2 and 3.

36. For a more detailed engagement of Frankfurt's work, see Fisch and Benbaji 2011, chap. 8, from which the following is partly distilled.

37. Korsgaard 1996a, 99n8, and 1996b, 394n34.

38. Taylor 1977, 15–16.

39. The citation referred to in the previous note is the only reference to Frankfurt I have come across in Taylor's writing. Korsgaard (2006), in contrast, discusses Frankfurt's later work in some detail.

40. Frankfurt 2006, 6.

41. Frankfurt 2006, 10.

42. Or as Frankfurt puts it in response to Richard Moran in Buss and Overton 2002, 219: "It is Identification that indispensably constitutes the source and the ground of reasons. The idea of identification as self-constitutive of agency is expressed most vividly in Frankfurt 1987, 170: 'in making a decision by which he identifies with a desire [the person] *constitutes himself*'" (emphasis Frankfurt's).

43. This does not cast first-order deliberation as necessarily frivolous or unimportant, of course. Resisting an impulse to act in the face of danger or weighing the costs and benefits of a desired course of action can pose heart-wrenching dilemmas without involving second-order desires.

44. The idea permeates Korsgaard's writings on normativity, as it does Kant's on

morality and practical reason. For pertinent examples, see Korsgaard 1996, 97–98, citing chap. 3 of Kant 1998 (1785).

45. Frankfurt 2004, 50.

46. Frankfurt 2006, 33.

47. Frankfurt 2006, 42.

48. Frankfurt 1987, 164.

49. Frankfurt 2006, 51–52.

50. Herman 2002.

51. Buss and Overton 2002, 276–78.

52. According to Frankfurt, their givenness, however, does not exempt us from being wary and critical about what we love and care about. Frankfurt 2006, §16, is devoted to our obligation to "keeping an eye out for possible corrections of our views" because "there are plenty of ways that our loving can go absolutely wrong" (48). However, getting ourselves right, for Frankfurt, turns out not to be about determining the rightness or wrongness of loving *X*, but is limited (a) to correcting "our views concerning what is important to us because our love for one thing conflicts with our love for another"; (b) to properly understanding the things we love; and (c) to making sure that we haven't gotten ourselves wrong, and really do love *X*. For Frankfurt, too, the idea of keeping the source of our norms in *normative* check does not even arise.

53. See especially McDowell 1978 and 1979.

54. See note 52, above.

55. McDowell 1994, 79–81; McDowell 1995; Fisch and Benbaji 2011, chap. 6, §1; and above, pp. 24–26.

56. See, in particular, McDowell 2009a, chaps. 14–16, 18, and 19; and 2009b, chap. 6.

57. Byrne 1996.

58. See, for instance, McDowell 2009a, 243–44. As noted above more than once, McDowell, like Brandom, rightly frames this as a question about the possibility of empirical *content*, rather than as an epistemological concern about empirical *warrant*. Framing it as a question about *intentionality* and *meaning*, rather than as a question of *truth* and *justification*, renders it crucial to the self-*critical* nature of empirical accountability, rather than to the self-*assuring* nature of empirical attestation.

59. McDowell 2009a, 244–45.

60. E.g., McDowell 1994, 68.

61. McDowell 2009a, 249.

62. McDowell 2009b, 6.

63. E.g., McDowell 1994, 94–95, 178. Brandom (2009, 43) stresses how the "relations of material incompatibility and inferential consequence" that constitute a "necessary condition of synthesizing a rational unity of apperception" by means of our linguistic repertoire "already implicitly involve commitment concerning the identity and individuation of *objects* they can accordingly be understood as representing or being *about*." Brandom's notion of the objects of judgment is analytical, however, and has nothing to do with the role of *bildung* in perceptual experience.

64. Toward the end of part 1 of Brandom 2009, Brandom makes the same point about perceptual judgments being constrained and "their evolution . . . subject to friction" by the "normative attitudes" that "practitioners are trained to acquire . . . non-inferentially"

(96). "Under the right circumstances," he explains, "properly trained observers are reliably disposed to respond to perceptible states of affairs by acknowledging commitments to corresponding perceptual judgments." But he steps more cautiously than McDowell in adding immediately that they are not granted special ontological or normative status by being immediate. In fact, unlike McDowell, according to Brandom that adjustment could go either way: "Like any other judgments, immediate perceptual judgments amount to petitions for recognition. The authority they claim may or may not be recognized by being incorporated in later rational integrations. But they exert constraint or friction just by making that petition for recognition."

Brandom, as we shall see shortly, locates the main source of normative friction, not in the "perceptual states of affairs" to which practitioners acknowledge commitment, but in the appraisal of other practitioners—an option barred by McDowell's insistence on interpreting the intersubjective turn in the fourth chapter of Hegel's *Phenomenology of Spirit* as an allegory of intrasubjective deliberation. See especially McDowell 2006, 2007, and 2009b, 185–203; and Pippin 2007 and 2011, 12–14.

65. Except to remark that, but never explaining *how*, self-critical ethical reflection can be "radical" to the extent that "one can find oneself called on to jettison parts of one's inherited ways of thinking" because "weaknesses that reflection discloses in inherited ways of thinking can dictate the formation of new concepts and conceptions" (McDowell 1994, 81).

66. McDowell 2009b, 105–6.

67. McDowell 2009b, 105–6. See the critique of McDowellian second nature, e.g., in Pippin 2007, 64–66.

68. And yet, as the quietism of *Mind and World* is gradually dispelled in later work, McDowell's position appears more and more to resemble the framework dependency characteristic of Kant's original picture, in which the normative shoe is located on the foot of spontaneity, rather than on that of the understanding. The framework to which he implies we are truly committed is the internalized, subliminal, tacitly applied framework of second nature, which, as in Kant's First Critique, the aim of science is to gradually make explicit, rather than to replace or improve. This is why the frameworks of science are answerable to second nature rather than the reverse. If this is true—and McDowell has yet to admit it explicitly—then his account is rendered far less relevant to the present study than I have implied.

69. I am referring to the first of the three chapters that make up part 1 of Brandom 2009.

70. Brandom 2009, 14.

71. The extension to practical reasoning is important, because throughout his earlier work, Brandom limits his discussion of normativity to *conceptual* norms. Asides like this one (see also Brandom 2009, 15, 35–36, 40, and 47) support attempts such as Stout 2002 and 2004, chap.12; Fisch and Benbaji 2011, chap. 6; and that of the present work, to extend Brandom's "normative pragmatism" to normativity in general.

72. Compare Kant 1965 (1781), A105 and A108.

73. Brandom 2009, 36 and 66–67.

74. Among the four thinkers discussed earlier, Frankfurt is the only one to voice a concern similar to Brandom's holism. His concern, however, as mentioned in note 52,

above, and discussed in greater detail in the next chapter, is primarily with resolving normative *ambivalences* with respect to one's objects of love, and has little to do with *inferential* consistency as such.

75. Brandom 2009, 15–16.

76. Brandom 2009, 15–16.

77. Frankfurt 1982, 182.

CHAPTER THREE

1. See especially Friedman 2001, 2002a, 2010a, and 2010b. Friedman's position is discussed and criticized in chapter 5 of Fisch and Benbaji 2011; it has also been commented and enlarged upon in detail by an impressive array of historians of physics and philosophers of science in Domski and Dickson 2010. Of course, relativizing the synthetic a priori is not confined to neo-Kantian accounts of science. Brandom makes the case more generally in an attempt to locate in, and then pry free from, "Kant's central categories" the "essential individuating feature of the metaphysical categorical sortal metaconcept '*subject*,'" namely, the a priori "metaproperty of normatively repelling" incompatible commitments—that is to say, the a priori conditions of criticizability itself (Brandom 2009, 47–49). However, as he himself admits, doing so diverges sharply from the framework of the First Critique in (a) *not* having "to presuppose the possibility of something called 'synthetic knowledge a priori,' and then search for the conditions of its possibility" and in (b) *having* to presuppose that "it must already be settled, at each stage of the process of rational critical and ampliative integration, what relations of material incompatibility and inferential consequence, the conceptual contents that are to be integrated stand in to one another" (47). In other words, Brandom agrees (if I am not missing something vital) that with regard to what grounds conceptual content itself (as opposed to the rules of synthesizing conceptual contents), Kant's original notion of the fixed universal constitutive a priori is replaced by a contingent, changing, a priori vocabulary, which, as we have seen in the previous chapter, does not add up, at least at the intrasubjective level, to a constitutive framework at all.

2. Lipton 2001. According to Richard Creath (2010, 508n1), Kuhn, in "The Road since *Structure*," his presidential address to the biennial meeting of the Philosophy of Science Association in October 1990, used the phrase "Kantianism with movable categories" to describe his position at the time. The phrase itself, claims Creath, was dropped in the printed version, but the idea, with its explicit reference to Kant, is clearly expressed: "the position I'm developing," writes Kuhn of his "theory of the lexicon,"

> is a sort of post-Darwinian Kantianism. Like the Kantian categories, the lexicon supplies preconditions of possible experience. But the lexical categories, unlike their Kantian forebears, can and do change, both in time and with the passage from one community to another. . . . Underlying all these processes of differentiation and change, there must, of course, be something permanent, fixed and stable. But like Kant's *Ding an sich*, it is ineffable, undescribable, undiscussible. (Kuhn 1990, 104)

See also Friedman 2010a, 712.

3. As I shall explain shortly, reluctance to run the two principles together owes to the profound difficulty of applying the second to the first in attempting to account for

the rationality of framework *replacement*—a problem that has been raised acutely only with respect to science. In other areas of philosophy, the question has not been raised, not because reasoned framework transitions are better understood there than in science, but because they are simply not discussed! As we have seen in the previous chapter, avid neo-Kantians, such as McDowell, Korsgaard, and Brandom, are quite comfortable running the two principles together, exactly because their work focuses exclusively on reasoning of the sort analogous to the kind of scientific deliberation Kuhn dubbed "normal."

4. Popper 1976 and 1994; also Bartley 1984, as explained in chapter 2, note 10, above.

5. Sellars 1956.

6. Which cannot apply to Bartley 1984, who purports to deny Framework Dependency quite generally, not only in science.

7. As noted in the previous chapter, much of mathematics is pursued in this way.

8. And to repeat the caveat with which I opened the previous chapter, the philosophical argument of this book should by no means be taken as an apologetic attempt to justify the claim that scientific framework transitions are always or mostly rational. My objective is to explore the *possibility* of such a transition and to articulate its structure in sufficient detail to be tested against the historical record.

9. An interesting early exception is Hilary Putnam's notion of revisable "framework principles," developed during the 1960s and 1970s in response to Quine's two dogmas argument. Responding to Quine, Putnam (1976) stresses the necessity of such principles—that is, their immunity to empirical refutation relative to the framework. The defining feature of such statements, he writes several years later, is that they "cannot be overthrown merely by observations, but can only be overthrown by thinking of a whole body of alternative theory as well" (Putnam 1994, 251). For neo-Kantians like Friedman and myself, the defining feature of scientific frameworks is not their immunity to refutation by empirical finding *per se*, but their being *constitutive* of such findings. Friedman's relativized a priori principles do not merely stand aloof from, and untouched by, the facts; rather, it is by them that the facts' very facticity is normatively established. Therefore, transitions from one scientific framework to another represent for Friedman, far more than for Putnam, a replacement of scientific *standards* that appears so profound as to defy reason. For a detailed account of the stable and the changing elements in Putnam's work on these topics, see Mueller and Fine 2005, especially §III; for a comparison of Putnam's and Friedman's positions in this respect, see Tsou 2010.

Among those who concede Friedman's approach, Robert DiSalle (2002, 2006, and 2010) is the only writer to date to insist on going beyond Friedman in describing the emergence of new constitutive coordinative definitions as owing to a decidedly critical "process of conceptual analysis," by which "internal difficulties" that plague the old are shown to "be resolved by the new." I discuss DiSalle's position further down.

10. The impressive attempt of Friedman (2001) to formulate a neo-Kantian account of the framework dependency of mathematical physics even keener than Kuhn's, yet which, unlike Kuhn's, boasts a robust notion of the rationality of framework replacement, is discussed extensively by Fisch and Benbaji 2011, chap. 5, and criticized (a) for concentrating exclusively on what I here call the second phase of the process, and (b) for lacking any critical element. I shall not rehearse that critique in what follows, although it will be apparent in the way the account I propose will build on and away from Friedman's.

11. As mentioned in chapter 2, note 11, above, the account of normative self-criticism presented in Fisch and Benbaji 2011, on which the present work builds, premises the contention, argued for there in some detail, that sufficiently diverse normative outlooks are normatively unrankable. The argument turns on the semantics of thin normative concepts like "good," "appropriate," and "deplorable," and is therefore quite general. Because the meaning of such concepts for a given community derives from the particular vocabulary of thick normative concepts it employs, sufficiently diverse normative outlooks lend them radically different meanings. (Think of how differently words like "good" and "deplorable" are understood in, say, decidedly conservative and decidedly liberal communities.) Because thin concepts mean different things for sufficiently diverse communities, no set of such concepts can exist that is capable of normatively ranking such outlooks in a manner acceptable to both. Put differently, for lack of such concepts, it is impossible for members of a community to deem their own outlook normatively inferior to a radically different one. It follows that if normative self-criticism is possible, it *cannot* be comparative. Friedman does not the address the problem explicitly, but seems to sense it in insisting that the differences between successive scientific frameworks are far less radical than Kuhn maintained, and hence are normatively comparable. See, for instance, Friedman 2001, 57–68, and the passage from Friedman 2010b referred to in note 16, below. See also remarks to this effect in Richardson 2002, 263.

12. Two of Friedman's central examples are the role played by the work of Mach on space and time, and of Helmholtz and Poincaré on the foundations of geometry, in Einstein's theory of relativity. See most recently Friedman 2002b and 2010b, 499–501.

13. Friedman 2010a, 714.

14. As DiSalle (2010, 527) puts it: these, according to Friedman, "provide physicists with . . . the occasion for rational and self-conscious reflection on the fundamental concepts of the framework, and the conceptual resources to think of [e.g.] otherwise bizarre Einsteinian notions, such as relative simultaneity and space time curvature, as intelligible theoretical possibilities." This, he justly submits elsewhere, is, "from my point of view, the most important part of [Friedman's] account of the rationality of science" (DiSalle 2002, 194).

15. Friedman 2010a, 714 (italics added).

16. Friedman 2010b, 500.

17. Friedman 2010b, 500–501.

18. See note 14, above.

19. DiSalle 2002, 192.

20. Notably DiSalle 2002, 2006, and 2010.

21. DiSalle 2002, 204.

22. DiSalle 2010, 528.

23. DiSalle 2010, 527.

24. DiSalle 2010, 527–28.

25. See Friedman 2010a, 718–29, and the first sections of chapter 4, below.

26. To argue thus is to go a step beyond DiSalle in the direction his important critique of Friedman points, namely, to insist not merely that in the course of paradigm shifts creative practitioners can be shown to have actively criticized components of the

prevailing framework, but that to be deemed rational, as I have argued from the outset, they are *required* to do so.

27. Rorty 1989, 13–14.

28. Rorty 1989, 20. This is indeed the position Rorty presents in great detail in *Contingency, Irony, and Solidarity* (1989). He grounds it explicitly and centrally in Bloom's notion of the "strong poet," mentioned above, who is motivated by an "anxiety of influence" to experiment ironically with final vocabularies different from her own with the hope of hitting upon a worthwhile alternative. Rorty's more serious work, however, such as his pointed critique of the history of epistemology in *Philosophy and the Mirror of Nature* (1979), is certainly far more a critical engagement of the philosophical tradition than an ironic casting around for new alternatives.

29. For example, DiSalle 2002, 194, and 2010, 526; but, as noted, without taking due note of the philosophical difficulty involved.

30. For a detailed critique of *Dynamics of Reason*, see Fisch and Benbaji 2011, chap. 5.

31. The problem of the source of rational incentive and capacity for finding one's norms normatively wanting has nothing to do with whether or not conceptual resources for subsequently framing a new framework are locatable ready-made beyond the confines of one's community. If the problem of rational incentive can be solved, as I shall suggest it can, the problem of conceptual resources loses much of its urgency (as DiSalle's criticism of Friedman clearly implies). And if it cannot be solved, explaining transitions by appeal to the mere availability of such resources, as Friedman does, steers too close to Rortian irony for comfort.

32. See especially Carnap 1937, §50–52, 82.

33. I am referring to the paragraphs of *Structure* that John Earman deems "purple." See Kuhn 1970, 94, 111–14, 150, 151; and Earman 1993, 19. The view is forcefully restated in Kuhn's "Postscript—1969" appended to the 1970 edition, where he concludes: "The conversion experience that I likened to a gestalt switch remains therefore at the heart of the revolutionary process" (1970, 204).

34. For one of his most pointed statements to this effect, see Kuhn 1993, 251–52.

35. Friedman 2001, 51–53, 83–85, 93–103.

36. Fisch and Benbaji 2011, chap. 5.

37. The problem should not be confused with that of the justificationary regress that Pippin, for example, makes much of in his discussion of the reflective, normative grounding of Kantian self-legislation. Pippin points vividly to the incoherence of "the image of some putatively law-less person making or originating or legislating a principle and only thereby being bound to it," for one must ask "on what sort of basis such a law-less subject could decide what to legislate." "Unless you are already bound by the constraint of reason, on what basis could you subject yourself to such constraints?" (Pippin 2007, 70–71). The problem addressed here is not that of the normative grounds for deeming one's norms binding, but that of the normative grounds for deeming them normatively *wanting*. Pippin (2011, 22–23) deems "ridiculous" attempts to ascribe to Kant or Hegel the idea that claiming that "persons are subject to no law or norm other than ones they have subjected themselves to" necessarily presupposes "episodes of self-subjection or explicit acts of allegiance." All it means, he argues, is that they are accepted by being followed—

subjects "follow the rules, are not governed by them." However, Pippin goes on to claim that "it is only because of this that someone like Socrates or Galileo or Freud can occasion intellectual crises." The fact that we are capable of violating the laws and norms we "give ourselves" is indeed a necessary condition for the very possibility of such "crises," but to imply that it explains the normative grounds by which such individuals are able to adopt such a self-critical stance begs the question this study purports to address.

38. This is the idea that grounds Hattiangadi 1978 and 1979, the most comprehensive analysis of the nature of problems to emerge from the Popperian camp. Several thinkers, including myself, have criticized Hattiangadi for narrowing problemhood to logical inconsistency. Giunti (1988) criticizes his account for its inability to account for problems that are not inconsistencies, such as those of measurement accuracy and gaps in knowledge. I have criticized him (Fisch 1994a and 1997, chap. 1) for failing to recognize system malfunction as paradigmatic of problemhood, and failing to distinguish problems large from small. But even I at the time failed to recognize the essential framework dependency of my own suggestions. The real problem with Hattiangadi's account is its inability to distinguish between formal incompatibility and inappropriateness.

39. The nonvoluntary "necessities of a person's will," writes Frankfurt (2004, 50), to quote at greater length a passage cited briefly in the previous chapter,

> guide and limit his agency. They determine what he may be willing to do, what he cannot help doing, and what he cannot bring himself to do. They determine as well what he may be willing to accept as a reason for acting, what he cannot help considering to be a reason for acting, and what he cannot bring himself to count a reason for acting. In these ways, they set the boundaries of his practical life; and thus they fix his shape as an active being.

40. Brandom 2009, 64 (emphasis added).

41. See chapter 1, note 20, above.

42. Briefly described in my introduction, above.

43. "The rationality that qualifies us as sapiens . . . can be *identified with* being a player in the social, implicitly normative game of offering and assessing, producing and consuming, reasons" (Brandom 2000a, 81; emphasis added).

44. Brandom 2002, 13–14. The common-law judge example is further developed in Brandom 2009, 84–90.

45. Fisch and Benbaji 2011, chap. 7, §1.

46. The eight historical chapters comprising part 2 of *Tales of the Mighty Dead* are candidly described by Brandom as a series of self-serving exercises of "genealogical, historical, expressively progressive reconstructive rationality," devoted, in one sense, to "rewriting the history of philosophy to make the present day a safe and congenial environment for views that are in any case going to be recommended" (Brandom 2002, 15–16). On Brandom's (and Heidegger's) "tendency to treat their predecessors as partial and sometimes confused versions of themselves," see Haugeland 2005 (quotation on 421); and more recently on the philosophical appropriation (or in his word "interanimation") of past thinkers in the hands of contemporary philosophers in general with special reference to Brandom's case, see Pippin 2015.

47. Brandom 2009, 85.

48. And in so doing, could be described as perhaps exemplifying the thesis presented in this chapter and the next—certainly with respect to my own thinking.

49. As opposed to normative incommensurability, or as Tsou (2003) has it, "the incommensurability of values."

50. Unless, of course, endorsing the new requires no more than to supplement the old, as when accepting additional responsibilities that have no effect on one's former obligations. But with regard to the kind of scientific framework transitions under discussion, this is never the case.

51. Wittgenstein (1974, 81e). The entire passage reads as follows:

> 608. Is it wrong for me to be guided in my actions by the propositions of physics? Am I to say I have no good ground for doing so? Isn't this precisely what we call a 'good ground'?
>
> 609. Supposing we met people who didn't regard that as a telling reason. Now, how do we imagine this? Instead of the physicist, they consult an oracle. (And for that we consider them primitive.) Is it wrong for them to consult an oracle and be guided by it?—If we call this "wrong" aren't we using our language-game as a base from which to *combat* theirs?
>
> 610. Are we right or wrong to combat it? Of course, there are all sorts of slogans which will be used to support our proceedings.
>
> 611. Where two principles really do meet which cannot be reconciled with one another, then each man declares the other a fool and heretic.
>
> 612. I said I would 'combat' the other man,—but wouldn't I give him *reasons*? Certainly; but how far do they go? At the end of reasons comes *persuasion*. (Think what happens when missionaries convert natives.)

52. The following few pages briefly distill part of the relatively detailed phenomenology of critical engagement presented in chapters 7 and 8 of Fisch and Benbaji 2011. By combining a Brandomian approach to normative discourse, a Walzerian view of social criticism, and a Frankfurtian account of personal identity, we are able to locate criticism, and especially normative criticism, as straddling the intersubjective/intrasubjective divide in a manner irreducible exclusively to either side. To understand the transformative effect of normative criticism is to understand not only of what such criticism consists and how it is *produced*, but also how it is *consumed*. As briefly noted above in the introduction, thinkers like Brandom and Walzer invest much in accounting for the social settings of the former, but have nothing to offer regarding the latter, namely, the impact of such discourse on individual commitment. Thinkers like Korsgaard, Taylor, Frankfurt, and to an extent McDowell have much to say about personal identity and the sources of personal normative commitment, as we have seen, but they do so *in vacuo*, giving no account of the self's capacity to surpass the limits of its personal normative commitments when, as lively Brandomian participants in the game of giving and asking for reasons, they find themselves subjected to keen Walzerian critics.

53. Fisch 1994a and 1997, chap. 1.

54. For an intriguing critique and complexification of the means-ends relation and its bearing on the logic of instrumental reason, see Frankfurt 1992.

55. For a discussion and comparison of the two positions in the political realm, see Fisch 2003.

56. Neither do the problems brought to the fore in prudent normative criticism resemble the sort of anomalies of which Kuhn speaks, which are recognized as problematic by members of the community and set aside. Once deemed worthy and adopted, a wholeheartedly applied norm or standard cannot give rise to conduct that could be considered anomalous from *that* perspective.

57. See, for example, his discussion of Locke's use of "constitution" in Skinner 1974, 285–87.

58. Frankfurt 1987, 163.

59. Frankfurt 2004, 92.

60. Frankfurt 2006, 95.

61. Frankfurt 1994, 139.

62. Frankfurt 2004, 91.

63. Frankfurt 2004, 95.

64. In his otherwise laudatory review of Frankfurt 2004, Quinn (2004) objects to Frankfurt's unyieldingly negative view of ambivalence. "Our lives would be impoverished," he writes, "if we did not allow room in them for a measure of restless striving toward a kind of self-transcendence that is bound to put the unity of the self at risk." When, as proposed here, Frankfurt's picture is enriched in the direction indicated by Quinn, our lives are rendered not merely more meaningful, but more *rational*!

65. For the discussion in this and the next two paragraphs, I am grateful to questions put to me by one of the two anonymous readers for the press.

66. Work I have done on the transformative impact William Whewell's early career as a textbook writer had on his philosophical understanding of science is an interesting example in this respect. See Fisch 1991a, chap. 2, and 1991b, §3.

67. Which might explain why Whewell, the young Trinity don, began to awaken to the problems afloat only from his third textbook onward . . .

68. The Talmudic literature is the only record of an extended scholarly undertaking that I am aware of that explicitly makes the difference between disputing an imaginary and a real critic the cornerstone of its approach. I am referring to the foundational story of the mythical heavenly ruling in favor of the Hillelite approach to deciding the law in the dispute with the Shammaites. For a close reading of it along these lines, see Fisch 2017.

CHAPTER FOUR

1. DiSalle would seem the exception: following Friedman in conceding Framework Dependency, while insisting, contrary to Friedman, that framework transitions in mathematical physics owed to their practitioners' criticism from within. But in so arguing, DiSalle reveals his commitment to a considerably weaker version of Framework Dependency than presupposed here (and possibly also by Friedman).

2. See especially Fisch and Benbaji 2011, chap. 5.

3. Agassi 1964 and Berkson 1995, discussed above in chapter 1, were mentioned as pertinent examples in this respect, as was Lakatos 1970, discussed in chapter 2.

4. Brandom 2009, 79.

5. Brandom 2009, 79.

6. Brandom 2009, 82–84.

7. Friedman 2001, 58. For my arguments against his use of the term, see Fisch and Benbaji 2011, chap. 5, §4.

8. Which, as we shall see, is far more a departure from the Kuhn of *Structure* than from the Kuhn of the study of Planck in Kuhn 1987.

9. I take the liberty of combining the two proposals because both Friedman and DiSalle view the latter's as complementing the former's. As Friedman puts it:

> DiSalle takes his approach to be complementary rather than opposed to my own: he is adding something to my version of the dynamics of reason rather than contradicting it. I view his approach, similarly, as complementary rather than opposed to mine, and I have no doubt, in particular, that critical conceptual analyses of the kind DiSalle describes in fact play a central role in the development from Newton to Einstein. (Friedman 2010, 718)

10. See especially Galison 1997, chap. 9.

11. Friedman 2001, part 2, chap. 4. In later works the terms "meta-framework" and "meta-paradigm" are dropped in favor of the less misleading "meta-scientific level"—as in Friedman 2010, 718.

12. Galison 1997, 784.

13. "Thus, for historical reasons, instead of searching for a positivist central metaphor grounded in observation, or an antipositivist central metaphor grounded in theory, I suggest that we admit a wider class of periodization schemes, in which three (or more) levels are intercalated" (Galison 1997, 799). Indeed, Galison's arguments against the two "central metaphors," as well as those he mounts in favor of his own, are predominantly historical rather than philosophical.

14. Galison 1997, 798.

15. Galison 1997, 798.

16. However, at critical moments, Galison's text veers toward a far more radical philosophical stand against Kuhnian antipositivism, as in his brief attack on incommensurability between theoreticians (Harman and Galison 2008, discussed briefly in note 28, below), and his more sustained philosophical self-description in the last section of Galison 1997 as combining "a historicized neo-Kantianism" (described [842] in terms similar to Friedman's notion, late of Reichenbach, of a relativized a priori) with "the justly influential work of Donald Davidson and Hilary Putnam." Even there, however, he describes his project as "manifestly more historical" (840) than philosophical.

17. Galison 1997, 803.

18. Friedman 2001, 105.

19. Galison 1997, 783.

20. Galison 1997, 803.

21. Notably Gorman 2002 and Collins et al. 2007.

22. Harman and Galison 2008, 568–69.

23. For a detailed account of the transformative role such didactical deliberations played in the formation of William Whewell's understanding of mathematical physics, see Fisch 1991a, chap. 2.

24. For a usefully nuanced typology of all such interactions, see Collins et al. 2007.

25. Fisch and Benbaji 2011, chap. 5, §5.

26. Galison 1997, 833.

27. Galison 1997, 830–31.

28. In Harman and Galison 2008, 568, Galison very briefly applies the trading-zone metaphor within a single subculture of physics rather than between two subcultures. His example is the way Poincaré and Lorentz engaged each other's work. But his comments are far too brief to be instructive. He does not clarify what actual trading they conducted, the nature of the pidgin or creole they employed, or how their views were transformed by the exchange, if at all. If all he purports to claim is that despite their differences (which remained in place throughout), they simply "did not, in fact, meet each other's work with gaping-mouth incomprehension," he is not claiming much. As Friedman argues convincingly, the problem of inter-framework discourse is not that of understanding, but that of convincing and transforming the thinking of practitioners committed differently.

29. Galison 1997, 51.

30. Galison 1997, 13.

31. Galison 1997, 51.

32. Galison 1997, 17–18.

33. Friedman purports to do so, at the price of returning to a more orthodox Kantian position (late of Cassirer), according to which science is such that, when properly conducted, its successive paradigms will display such "inter-paradigm convergence" (Friedman 2001, 66) as to render "our present scientific community . . . an approximation to an ideal limiting community of inquiry acknowledging similarly ideal standards of *universal, trans-historical* communicative rationality" (68; italics added). Yet again, even if a successive elaboration of humanly fashioned linguistic frameworks can achieve such transhistorical universality, all it will have achieved is universal human agreement, with no greater provable "purchase on 'truth' or the 'world,'" than any other linguistic framework. Under conditions of strict framework dependence, the very notion of truth by correspondence to an *independent* reality is inherently incoherent and cannot be considered even a regulative idea. Be that as it may, Friedman's solution is evidently not an option for Galison, who dissociates himself explicitly from the sense in which "Kant's 'transcendental' suggests conditions that are beyond history." "The entirety of the last eight chapters" of *Image and Logic*, he writes, "speaks against such an assumption" (Galison 1997, 842).

34. Galison 1997, 835; emphasis added.

35. Galison 1997, 834.

36. See Warwick's (2003) insightful study of the training of mathematical physicists in nineteenth-century Cambridge, especially chaps. 2–5; and Collins's equally insightful account of the taken-for-grantedness of a scientific community's basics in the section entitled "Wittgenstein and Rules" (Collins 1985, 12–16).

37. As when preparing a lecture or report, or writing a textbook. On the relation between real and effectively imagined encounters of this kind, see above, chapter 3, notes 65–68 and associated text.

38. George Peacock's deeply ambivalating engagement during the early 1820s with Charles Babbage's thoroughly formalistic approach to analysis provides a vivid illustration of the ambivalating impact talking across paradigm dividing lines is capable of hav-

ing. The story of the formation and subsequent impact of Peacock's *Treatise on Algebra* (1830) is related in full in the three chapters comprising part 3, below. For early versions of this first part of the story, see Fisch 1994a and especially 1999.

39. Friedman 2010b, 500.

40. Friedman 2001, 108.

41. Moreover, not only is the force of our convictions not weakened by knowing that most people hold very different views, it is not weakened even by fully realizing that had *we* been raised or trained differently, we would in all probability be committed just as deeply to very different norms. Or as Frankfurt would put it, *knowing* that our volitional necessities are ultimately the product of contingent circumstance does not make them less necessary for us. G. A. Cohen aptly calls this the "paradox of conviction"; see especially Cohen 2000, chap. 1.

42. On the difference between "incommensurability of meaning" and of "value" in Kuhn, see Tsou 2003 and the text to note 48 in chapter 3, above.

43. Haugeland 1998, 340–43.

44. As noted in chapter 2, Lakatos's historiography of scientific research programmes resembles the later Kuhn in subjecting transitions from one research programme to another to purely instrumental considerations of predictive fruitfulness. Commitment within a programme to the programme's "hard core" is normative, in the sense that it is rendered immune to empirical refutation by the ad hoc construction of "auxiliary hypotheses." Commitment to the programme itself, however, is, for Lakatos, a matter not of empirical testing, but of empirical *fertility*. For his methodology and historiography of scientific research programmes, see Lakatos 1971 and 1978. For recent expositions of Lakatos's position in comparison to Kuhn's and to Friedman's, see Gattei 2008, 56–62, and Koertge 2010, respectively. For references to criticisms of Lakatos, see above, chapter 1, note 6.

45. Whewell 1837b, the first-ever comprehensive history of science, set the tone with the fixed historiographical pattern it imposed on all scientific development. It viewed the history of every science as punctuated by a series of "inductive epochs," in which a novel idea was discovered successfully to "colligate" a body of formerly unconnected fact. Whewell's epochs are flanked on either side by "preludes" of work preparatory to the discovery and "sequels" in which its consequences were worked out in detail. For Whewell—as in virtually all history of science written to date—scientific development is conceived solely in terms of scientific *achievement*. Moments of indecisive dithering have no place at all in Whewell's narrative schema (and, consequently, find no place in the semi-Kantian philosophy of science he later articulated "founded on their history"). One self-conscious exception to the rule (other than Friedman) is the role given to intermediary figures in the historiography employed in Damerow et al. 2004.

46. Friedman 2001, 60.

47. Friedmn 2001, 61.

48. Friedman 2001, 60–63. Galileo's intermediary role lies in preserving the two basic Aristotelian forms of natural motion—naturally accelerated motion toward the center of the earth, and naturally uniform circular motion directed at right angles to it—while discarding "the hierarchically and teleologically organized Aristotelian universe" (61). According to Friedman, Galileo's notion of circular inertial motion rendered "the modern

conception of rectilinear natural inertial motion . . . actually continuous with the preceding Aristotelian conception of natural motion." (61). On Galileo as an intermediary figure, see also Damerow et al. 2004, chap. 3, especially §3.7.

To Poincaré's *Science and Hypothesis* (a book he maintains Einstein was "intensively reading . . . immediately before his revolutionary breakthroughs in 1905"), Friedman attributes the neo-Kantian idea that although geometry functions as "a constitutive framework making properly empirical discoveries first possible," the fact that "there is more than one such constitutive framework" can only mean that the choice between them is "based on a convention" (simplicity), and not on "some innate necessity of the human mind" (62). In this way, "Einstein's introduction of a radically new constitutive framework for space, time, and motion again grew naturally out of, and is thus quite continuous with, the preceding framework it replaced" (63).

49. Friedman 2001, 107.

50. Friedman 2001, 63.

51. Friedman 2001, 63.

52. Friedman 2001, 63–64.

53. A point, as we have seen, that is criticized pointedly, but unsatisfactorily, by DiSalle.

54. See note 48, above.

INTERLUDE

1. Nonetheless, as I have argued elsewhere, rationality in the latter, game-theoretic sense cannot dispense entirely with the former sense. See Fisch and Benbaji 2011, chap. 1, §1.

2. For example, Kuhn 1993, 338; firmly and convincingly contested by Friedman 2001, 83–84.

3. G. A. Cohen 2000, chap. 1.

4. For a similar, yet differently argued distinction see L. J. Cohen 1989, 1991, and especially 1992, on the differences between belief and acceptance.

5. See chapter 4, note 41, above.

6. For criticism of Friedman along these lines, see Fisch and Benbaji 2011, chap. 5, §5.

CHAPTER FIVE

1. Herschel and Babbage 1813, iv. On the other founding members and their backgrounds, see Enros 1979, 103–22.

2. The fact that by the second decade of the nineteenth century Cambridge, and with it British mathematics in general, had fallen drastically behind the level of mathematics taught and practiced on the continent is undisputed. But whether or not it should be attributed to inherent inadequacies of the Newtonian methods and notation taught at Cambridge, as the Cambridge reformers and most historians—including myself—maintained, has been recently challenged by Grabiner 1997 and 2002, and especially by Guicciardini 2004.

3. On the Analytical Society see Enros 1979, 1981, and 1983; Becher 1980a and

1980b; Schweber 1981, 54–67; Guicciardini 1989, §9.5; Wilkes 1990; Fisch 1991a, 2.221–3, 1994c, and 1999; and Ashworth 2008.

Some disagreement exists regarding the date of the society's formation. Most historians (e.g., Enros, Becher, and Wilkes) follow Babbage's unpublished "History of the Origin and Progress of the Calculus of Functions during the Years 1809, 1810 . . . 1817" (otherwise known by its Oxford Museum of the History of Science call-name: Buxton MSS 13, referred to hereafter as MS 13) in asserting that the society was formed "early in May 1812," with the initial group first gathering in Bromhead's lodgings on "Thursday May the 7th" and holding the society's first official meeting the following Monday, May 11th (24–25). But there is evidence to the contrary: a letter from freshly graduated, newly appointed Fellow of Trinity John Brass to Herschel, dated 2 November 1811, addressing him as "President of the Analytical Society," declining the offer to join, but extending "warmest wishes for the success of a society which seems so highly calculated to contribute to the promotion of mathematical literature"—cited in Schweber 1981, 57–58. Enros (1983, 33), who with regard the Analytical Society gives priority to MS 13, emends the date of Brass's letter to 1812. Likewise, Hyman, who also adopts the spring 1812 dating, is forced to emend the date of a letter concerning the society sent by Maule to Babbage from 16 January 1812 to 16 January 1813. Indeed, in his own published account of the events, Babbage (1864, 20) states that the society was founded "at the commencement of my second year at Cambridge"—namely, autumn of 1811, rather than May 1812. Despite Babbage's vivid unpublished description, I am convinced that the earlier date is closer to the truth. The discrepancies between Babbage's published and unpublished accounts of his Cambridge days extend far beyond issues of dating, and give rise to serious doubts with respect to the validity of the latter. The former (Babbage 1864), as many have noted, is often unreliable, but, as I argue in detail in Fisch 2014, there is something disturbingly unreal and manipulative about MS 13, which, in the light of other evidence and its lack of coherence, should be largely rejected. It can certainly teach us a lot about Babbage's state of mind toward the end of his life, but should on no account be taken at face value as a reliable account of his early years.

4. Babbage's role during this phase of the story was crucial, as we shall see, but his career thereafter is less relevant to the concerns of the present study. After 1820 or so, he steered wide of the kind of meta-scientific deliberation in which I am interested—with the possible exception of an unpleasant rub with Whewell on the relevance of deductive, mathematical reasoning to natural theology in his self-appointed "Ninth Bridgewater Treatise" of 1837. Babbage remained unflinchingly faithful to a thoroughly formalist (or in latter-day parlance, algorithmic, or procedural) view of mathematics, forcefully formulated in his unpublished "Essays on the Philosophy of Analysis," written around 1820—a commitment evidenced in both his early work on the calculus of functions and his lifelong dedication to his difference and later analytical engines (see Grattan-Guinness 1992). Babbage, however, nowhere argued for, or even stated, such a position in print. I shall return to his "Philosophy of Analysis" in some detail below.

5. During the first decade of the century, sporadic protests lamenting the sorry state of British mathematics were voiced both within and outside of Cambridge, but to no avail (Guicciardini 1989, chaps. 7–9; Grattan-Guinness 1992, 35–36). The Edinburgh natural philosopher John Playfair (1808) and Cambridge-based Robert Woodhouse (1803, 1809,

and 1812) stand out in this respect. On Playfair, see Guicciardini 1989, §7.2 and 2004, §5; on Woodhouse, see Becher 1980b, Guicciardini 1989, §9.2, Phillips 2006, and references there.

6. For example, Warwick 2003, 66–68; and Guicciardini 2004. Grattan-Guinness (1992) is an exception in this regard. However, because his main focus is on Babbage (and follows Dubbey 1978 and Becher's [1980b] unfounded contention, discussed below, that Peacock's *Treatise* emulated Babbage's approach to analysis to the point of plagiarism), even he fails to appreciate the unsettling radicality of their program.

7. First noted in Fisch 1991a, 28–36, and argued further in Fisch 1999.

8. See note 2, above. In this respect, rendering the calculus analytic was akin to the act of reductive translation represented by Descartes's analytic geometry. Lagrange's version of the calculus went a radical *formalist* step beyond algebraization per se.

9. The first edition of Peacock's algebra was published in 1830, but the significantly modified second edition of 1845, as I shall argue in chapter 7, clearly militated against the radicalism of his initial hybridic account.

10. Guicciardini 1989.

11. Enros 1981 and 1983; Becher 1980a and 1980b; and Schweber 1981, 54–67.

12. Koppelman 1971/2; Richards 1980, 1987, 1991, and 2011a; and Pycior 1981, 1982, 1983 and 1984.

13. The Analytical Society is only mentioned in passing, and of the reforms it campaigned for, only the replacement of Newton's dots by Leibniz's *d* notation is noted. In her rich unfolding of the careers of the four friends on whom she focuses, Herschel, Babbage, and Whewell's early and insistent adoption of Lagrange's version of the calculus plays no role. See Snyder 2011, 32–35.

14. Forester to Whewell, 24 December 1841, cited in Todhunter 1876, 1:6; Snyder 2011, 19.

15. Snyder 2011, 35.

16. Cited by Buttmann 1970, 11, no source given.

17. Snyder 2011, 42–43.

18. Babbage 1864, 25. Snyder justly chides Babbage for taking credit in his autobiography for "every one of his actual achievements and several that are more dubious—for example, . . . even being the host of the meetings of the Philosophical Breakfast Club, when they were really held in Herschel's rooms" (Snyder 2011, 327). However, as noted, the passage in question does not take credit for anything related to the content or the membership that Snyder associates with her "Breakfast Club." Snyder insists that Babbage's claim to have hosted the Sunday breakfasts should be doubted, but offers no evidence to the contrary. However, whether they were held in Herschel's or Babbage's rooms is of little consequence to Snyder's story or to mine. For Babbage's complicated relationship with Herschel, especially in retrospect, see Fisch 2014.

19. On Babbage's "deep interest" in the philosophy of Bacon, see Buxton 1988, 351–53.

20. Snyder 2011, 37n56.

21. To Herschel.

22. Whewell 1859, xix–xx.

23. Schweber 1981, 17, by contrast, uses language very close to Snyder's in describing

the hopes of those who joined the Analytical Society as "nothing less than a transformation of British science."

24. It is also a story that, because so thoroughly informed by the *Novum Organon*, focuses almost entirely on issues of *methodology*, to the exclusion, again typical of Bacon's own writing, of the kind of historiographical and epistemological questions that, again, for Whewell at least, were eventually to become absolutely central.

25. Whewell's *Novum Organon Renovatum* (the name he gave to the second, philosophical part of the 1858 third edition of his *Philosophy of the Inductive Sciences*) stands to Bacon's *Novum Organon*, I have argued, as that work's title clearly suggested it stood to Aristotle's *Organon*—namely, as a *corrective*, rather than a continuation or elaboration. Whewell does not mince his words in declaring, on the first page of the preface to that work, that "as to the structure and operation of the *Organon* by which truth is to be collected from nature,—that is, the Methods by which science is to be promoted—we know that, though Bacon's general maxims are sagacious and animating, his particular precepts failed in his hands, and are now practically useless" (iii). This, as I have shown in detail in Fisch 1991a, was by no means a late development, but dates back to the mid-1830s.

26. A sporadic involvement that was often marred by disagreement with Herschel, Babbage, and especially Whewell. See Becher 2004.

27. Peacock was, however, deeply involved during the 1820s in an inductive investigation of algebra itself. His work on arithmetic, which began in a series of papers presented to the Cambridge Philosophical Society in the mid-1820s and culminated in his masterly entry "Arithmetic" in the *Encyclopaedia Metropolitana* of 1829, set forth self-consciously as "a natural history of arithmetic," to use Kevin Lambert's apt phrase. Peacock's 'philosophy of arithmetic' was grounded in a squarely empiricist conception of the idea of number that originally owed to Locke, and built on a meticulous and extraordinarily broad empirical survey of modes of numbering, counting, and calculating, past and present. Building on the work of Richards (1987 and 1991) and Durand-Richard (1990 and especially 2011), Kevin Lambert's (2013) detailed study of this phase in Peacock's development strongly suggests, as we shall see in chapter 6, that Peacock should be counted centrally among the members of Snyder's Baconian "Philosophical Breakfast Club."

28. Warwick 2003, 37.

29. Only after the establishment of the Moral Tripos and Natural Science Tripos in 1848 could one earn Cambridge honors without having first to take mathematics, and even then, only in 1854 were students reading classics fully exempt from mathematics! See McMackin-Garland 1980, chap. 7; Warwick 2003, 37; and Stray 2001; but compare Becher 1980a, 6, and Rothblatt 1968, 181.

30. For example, McMackin-Garland 1980, 30.

31. At the turn of the nineteenth century the Cambridge mathematics curriculum included arithmetic, algebra, geometry, trigonometry, fluxions, mechanics, hydrostatics, optics, and astronomy, culminating, in both its "pure" and "mixed" subjects, in Newton's *Principia* and *Opticks*. For additional details regarding the curriculum and the textbooks used at the time, see Gunther 1937; Rouse-Ball 1889, 119–21; Becher 1980a; and Warwick 1998, 2001, and especially 2003.

32. The term, coined by Henk Bos (2001, 10), forms the basis for Guicciardini's (2004) characterization of "the divide . . . [and] the lack of communication between Continental and British mathematicians, which set in during the second half of the eighteenth century" (248). See also Fraser 1990, 250–51.

33. Guicciardini 2004, 246. See also Fraser 1989 and references there.

34. Guicciardini 2004, 243.

35. See Playfair 1808, 264–65 and 323–24; Guicciardini 2004, 246–47; and Warwick 2003, 34.

36. Boyer 1968, 621.

37. Cannon 1964, 176–77. This is not to contest the more leveled assessments of Grabiner and Guicciardini.

38. Rothblatt 1968. See also McMackin Garland 1980, 18–19.

39. Buttmann 1970, 9; Snyder 2011, 28.

40. Buttmann 1970, 9.

41. Rogers to Herschel, 6 January 1812, Herschel Correspondence, Royal Society of London, cited in Ashworth 1996, 636.

42. Playfair 1808. For a detailed account of the review, see Guicciardini 2004, 246–48.

43. Lacroix 1802 and 1816, respectively. For a thorough evaluation of Lacroix's approach to the calculus, see Domingues 2008. For a detailed comparison between the three-volume *Traité* and the *Traité élémentaire*, see especially Domingues 2008, chap. 8.

44. Ashworth 1996, 636, citing Rogers to Herschel, 5 November 1808 and 6 January 1809.

45. Buttmann 1970, 9–10.

46. As noted above (note 3) Babbage compiled two contrasting accounts of his pre-Cambridge and Cambridge years—one published as chapters 3 and 4 of Babbage 1864, the other occupying the first hundred pages or so of his history of functional analysis, MS 13. For a detailed comparison between the two accounts and my reasons for generally rejecting the latter in favor of the former, see Fisch 2014. The incongruities between the two accounts are interesting in themselves, and in what follows I shall note in passing those that bear on the larger story the present study sets out to tell.

47. Campbell-Kelly 1994, 11.

48. Ditton 1706.

49. Agnesi 1801.

50. Woodhouse 1803.

51. Lagrange 1797.

52. Maclaurin 1742.

53. Presumably Simpson 1805.

54. Babbage 1864, 18–19.

55. Babbage 1864, 19.

56. Babbage 1864, 19.

57. Babbage 1864, 20.

58. Babbage 1864, 20.

59. Babbage 1864, 19. MS 13, by contrast, maintains that he purchased the book not in the fall of 1811, but in October 1810, en route to taking up residency at the outset of his *first* year. As I explain in Fisch 2014, the "great work of Lacroix" to which Babbage

refers was the three-volume *Traité du calcul différentiel et du calcul intégral*, and not the much shortened and simplified, one-volume *Traité élémentaire* of 1802. The *Traité* was "a monumental reference work: an encyclopedic appraisal of the calculus at the turn of the century," to quote Domingues's (2008, 23) authoritative description of the work. (Domingues interestingly points out how Lacroix subtly adopted as its motto the very same quotation from Horatio's *De arte poetica* that Diderot and D'Alembert had used as the motto of the *Encyclopédie*!) It was the kind of work that one uses either as a reference book or to structure a comprehensive course of studies. According to both accounts, Babbage bought it with the latter intention in mind. Hence the great discrepancy between them. According to his published autobiography (Babbage 1864), the decision to restructure his course of studies along the best continental lines was reached as a result of his frustrating first-year experiences at the university. His unpublished account of those years, MS 13, in contrast, portrays him fantastically as having arrived at Cambridge equipped with a clear sense of the deficiencies of the university curriculum, the vast superiority of continental analysis, and the singular merits of Lacroix, fully resolved from the very start to pursue an independent course of study. Everything we know about Babbage's childhood and early schooling militates forcefully against such a description.

60. At the "prohibitive price," as Dubbey (1978, 32) puts it, of seven guineas, to be precise.

61. See, for example, Schweber 1981, 57–63; and reference to Snyder 2011 in note 18, above.

62. Except for the highly misleading MS 13, that is.

63. MS 13 goes to great lengths to imply that he and Herschel first met only after the formation of the Analytical Society, which, according to that account, was in mid-May 1812, that is to say, weeks away from the completion of his second year. See Fisch 2014 for my reasons for also disbelieving this aspect of the story told there.

64. Babbage 1864, 20.

65. See note 16, above.

66. I am referring especially to Herschel 1813a, 1813b, 1813c, 1814, 1819a, and 1819b; and Babbage 1815, 1816, 1817, and 1820.

67. Babbage, as we shall see shortly, was eventually to abandon Lagrange's version for an interestingly more starkly formalistic approach of his own design.

68. The following pages sample, correct, and enlarge upon earlier work of mine related to the society in Fisch 1991a, 1994c, and 1999.

69. Babbage 1864, 21.

70. The preface presents an impressive survey of the history "up to the present time" of mathematical analysis, which it identifies with "symbolical reasoning" (ii), with special reference at the end to functional equations (xvi–xviii). For reasons explained in Fisch 2014, MS 13 goes to great lengths to attribute the initial drafting of the preface exclusively to Babbage, with later, brief comments by Herschel. But as Schweber (1981, 60–63) shows conclusively, the manuscript of the *Memoirs* in the archives of St. John's College, Cambridge, proves that this could not have been the case, and that Babbage's description of the preface as composed "Jointly with Sir John Herschel" in the "The List of Mr. Babbage's Printed Papers" appended to *Passages*, is closer to the truth.

71. An approach forcefully adopted by Enros 1979 and 1983. See note 74, below.

72. Herschel and Babbage 1813, iv.

73. Work published on the Analytical Society since Philip Enros's dissertation of 1979, especially by Schweber (1981) and myself, has tended to contest two of Enros's main claims. According to Enros (1979, 157), the Analytical Society disbanded in 1813 with the graduation of most of its members. It had existed, on his reckoning, for a period of just over a year; see also Grattan-Guinness 1992, 36. Enros also insists that, although dissatisfaction with the mathematics taught at the university "had been a factor" in the formation of the society, it "was not founded to change or reform the mathematical studies at Cambridge . . . but rather to further analysis" (1979, 121 and 165). He portrays the society primarily as "a mathematical society" that functioned to "aid in the mathematical research of some of its members" (165), and subsequently treats Herschel, Babbage, and Peacock's campaign to change things at the university as a later and distinct episode, for which the society "was the precursor" (Enros 1983, 26). In one obvious sense he is right: before obtaining college positions, members of the society could have very little direct influence on the curriculum—a task to which Peacock and later Whewell set themselves in earnest after their Trinity Fellowships had been secured. One way for a society of undergraduates to do so indirectly, from the outside, was by making available to the university a suitable English-language textbook. Enros ignores the fact that even Babbage, who would never have much to do with students or university teaching (he held the prestigious Lucasian Chair from 1828 to 1839, but as far as we know he never taught), not only claimed repeatedly that the need for an English-language edition of the smaller Lacroix had motivated the formation of the society in the first place, but that he himself had translated the first part, with Herschel and Peacock finishing the job and providing the notes. The three cofounders had also resolved early on to collaborate in the production of a compendium of problems, which although only published in 1820, was publicly promised in the advertisement to their edition of Lacroix in 1816. Had the society indeed disbanded in 1813, as Enros claims, he could still have argued that all of this belonged to a later phase, but apparently it did not. The most decisive piece of evidence against Enros's account are the minutes of the society's last known meeting, preserved in the British Museum as MS Add. 37182, fol. 64 (cited at length by Schweber 1981, 64)—a meeting reported to have been held at the house of Edwin Ffrench Bromhead in Thurlby near Lincoln on 20 December 1817! Further proof of the idea of the society being still very much alive in 1817 can be found in the latter part of a letter Herschel wrote to Babbage in January of that year:

> We must get another vol. of the Anal. Socy out, you must write a paper—so must Bromhead—we will ask Peacock, and get Ivory or Wallace to Present us a paper.—It will be capital sport.—What do you say? Will you take part—shall we revive the Anal. Soc. in a more moderate way, & in good *earnest* and get Ivory & those people for members—never mind the nonsense of having weekly or monthly meetings—an annual or quarterly dinner in Town will do just as well. (Herschel to Babbage, 30 January 1817, RS HS 20.38)

I stress all of this, because by ignoring the Society's protest and rebellious self-image, playing down its plans for reform (even though they could only have been realized later), and portraying it as geared primarily toward research, Peacock, who did not engage in original research during those early years, is virtually written out of Enros's account, and is relegated, by implication, to the role of an awestruck groupie (Enros 1979, 120). Content with an all-too-general definition of "analytics as implying "an algebraic or formal, operational approach to the subject matter" (5), and focusing exclusively on the original early mathematical work of Herschel and Babbage, Enros, by omitting Peacock's contribution, misses, we shall see, the important disagreements, both foundational and tactical, between him and Babbage evidenced in the latter's correspondence and notes to Lacroix.

74. Babbage 1864, 20.

75. At this juncture I am referring exclusively to the twelve notes penned by Peacock. The position adopted in Herschel's share of the notes, as we shall see shortly, was different, and closer to that of Babbage. Of course, neither did Herschel's notes in any way qualify Lacroix's original notation.

76. The account related in Babbage 1864 is borne out by the brief allusion to the formulation of Babbage's initial plan on page 24 of MS 13: "Early in May 1812 on the occasion of Mr. Slegg's mentioning the disputes which at the time existed relative to the distributing the bible alone or the bible accompanied with comments. I suggested the plan of instituting a society for the distribution of the small Lacroix" (the word *alone* is underlined in the original).

77. For an account of Lagrange's version of the calculus in comparison to the other main versions developed in (acknowledged or unacknowledged) response to Berkeley's 1734 pointed attack on the methods, conceptual structure, and logic of those originally proposed by both Newton and Leibniz, and his own outrageous alternative (that the calculus works effectively because of a systematic "compensation of errors"), see Fisch 1991a, §2.22, from which the present section acquires its name, and especially Guicciardini 2003, and references in both.

78. For Duhem's instrumentalism, see Duhem 1954; for his instrumentalist reading of Lagrange's mechanics, see Duhem 1980, chaps. 6–8.

79. Peacock to Babbage, 7 May 1822, London, British Museum, Add. MS 37182, No. 323.

80. London, British Museum, Add. MS 37202.

81. *Proceedings of the Cambridge Philosophical Society* 7 (1890): xx. He presented the former for Babbage in December 1825.

82. Folios 5 and 6 of the "Philosophy of Analysis" appear to be the draft of a cover letter to the Cambridge Philosophical Society, accompanying the paper he delivered in December 1821, that announces and outlines the entire project. In which case, it is doubtful that it, too, would have been included in the package to Peacock. Dubbey mistakenly describes it as a draft of the introduction to the projected book. The opening paragraph begins as follows: "In presenting to the Cambridge Philosophical Society the first of a series of *essays on the Philosophy of Analysis* some account of the object and nature of the discussions they will contain as well as a sketch of the path and purpose [they] pursue may render their intimate connections more evident." It is quite clear that the "Philosophy of Analysis" represents work in progress of which only the first essay, or chapter, had been completed by the time Peacock had read it. I doubt that Babbage, who quickly became engrossed in the difference engine project, ever worked much further on any of the papers. So in all likelihood the material Peacock had read and responded to was very close, if not identical, to the version of "Philosophy of Analysis" in hand.

83. Fisch 1994c and 1999.

84. Peacock seems to have produced no original mathematical work during the early years. Babbage concentrated for the most part on functional equations, and Herschel on the calculus of finite differences and their bearing on functional equations. Detailed assessments and comparisons of their early papers can be found in Schweber 1981; Enros 1979, chap. 5; and Koppelman 1971/2.

85. Schweber 1981, 59–60, citing W. H. Maule to Babbage, 16 January 1812, British Museum Add. MS 37182, fol. 3.

86. Dubbey 1978, 93–107; a view wholly endorsed by Hyman (1978) though toned down somewhat in Hyman 1982, 36.

87. Hamilton to Peacock, 23 July 1835, Trinity College Library, Add.ms.b.4944.

88. For criticism of Hamilton's taxonomy, see Fisch 1994c, 258–59. Despite the flaws, Ivor Gratten-Guinness (1996, 359) makes good use of it in analyzing the nature of Euclid's *Elements*, though somewhat differently from the way I believe Hamilton intended his distinctions. The advantage of Hamilton's taxonomy is that it more clearly marks off what I here term "formalistic" from other analytical approaches. In particular, "analytical" was generally used, then and now, as synonymous to "algebraic" (as opposed to the synthetic methods of geometry), thus glossing the differences between different accounts of algebra itself. In this regard, see especially Bos 2001; and Guicciardini 2004.

89. Hamilton 1837, 293.

90. Hamilton 1837, 294.

91. Ten years or so later, his work on quaternions and symbolic geometry well on its way, Hamilton seems to have softened to the point of defying his original terminology. In the course of acknowledging the existence not only of people who are drawn to algebra's philological difficulties, but of a sort of "*symbolical science*, or *science of language* which well deserves to be studied . . . [in which] *forms of expression* [are] treated as themselves the subject-matter to be studied," he admits to feeling "an increased sympathy with, and fancy that I better understand that Philological School" (quoted in Graves 1885, 2:521–52). See also Hamilton 1853, 14–16; and Koppelman 1971/2, 226.

92. Another good reason for casting Peacock's story in terms of Hamilton's taxonomy of "algebraists" is that it offers something of an insider's view of the story itself in explicitly informing Hamilton's (mis)reading of the *Treatise*, and motivating his important response to it that is taken up in chapter 7.

93. DeMorgan 1835, 311.

94. "Essays on the Philosophy of Analysis," fol. 42.

95. A view that, to the best of my knowledge, was first laid out in writing in a letter to Bromhead of 6 August 1815, now in the private collection of Sir Edward Ffrench Bromhead's correspondence in the possession of Sir Benjamin Denis Gonville Bromhead, Thurlby Hall, Thurlby, Lincolnshire (cited in Enros 1979, 179).

96. An essay entitled "<Of> General Notions Respecting Analysis," "Philosophy of Analysis", fols. 41–55; quotation at fol. 44.

97. As we shall see, Babbage was not merely proposing *a* calculus of operations, but was redefining algebra itself as a wholly abstract, nonquantitative calculus. Working almost exclusively from published material, standard histories of the development of abstract algebra in England, such as Clock 1964, Koppelman 1971/2, and Novy 1973,

have completely missed this early contribution, which in view of Peacock's acquaintance with it is of crucial importance. Though perhaps not fully appreciative of the meta-mathematical import of Babbage's proposals, Enros (1979), by contrast, takes full account of its existence. Unfortunately, for reasons explained in note 74 above, Enros chose to leave Peacock out of the story, and as a result, despite his superior appreciation of Babbage's unpublished work, he misses its one real impact.

98. Koppelman 1971/2, 187; but compare Enros 1979, 179.

99. The general plan to found "the whole system on the principle of identity" was apparently formulated and discussed with friends in some detail as early as the summer of 1815; see note 95, above, and Babbage 1869, 202. Enros (1979, 202n2) draws attention in this regard to a misfiled letter from Babbage to Brewster of 20 June 1821 (British Library, Babbage MS), which I have so far been unable to consult.

100. Babbage, MS 13, p. 217.

101. "Philosophy of Analysis," fol. 44.

102. As we shall see in chapter 7, below, Whewell conceived of a "pure" science, not unlike Hamilton, as a well-formed, axiomaticized, "Euclidean," deductively connected scheme of propositions derivable from a set of true axioms. The notion of axiomatic truth is central to his system. Axioms of a "pure" science (or the pure component of an "inductive" science) are intuitively recognized as stating true features of a "Fundamental Idea." They do not reduce to arbitrary definitions, but to basic truths concerning basic ideas. For an earlier account of Whewell's theory of the truth of "pure" science, see Fisch 1991a, 154–59. For his unsuccessful attempt to include symbolical algebra within this scheme, see Fisch 1994c, 266–76.

103. Koppelman 1971/2, 157.

104. DeMorgan 1842a, 288–89.

105. "Philosophy of Analysis," fol. 44.

106. "Philosophy of Analysis," fol. 46–48 (my italics).

107. This is not to say that one should resist generalizing indexicals in all cases. Babbage and Herschel were among the first in England to take the analogy between

$$x^n \cdot x^m = x^{n+m};\ \frac{d^n}{dx^n}\left(\frac{d^m u}{dx^m}\right) = \frac{d^{n+m}u}{dx^{n+m}};\ f^n\left(f^m(x)\right) = f^{n+m}(x)$$

seriously enough to seriously consider cases of the latter in which n and m are fractional or imaginary. Herschel in his contribution to the *Memoirs of the Analytical Society* allowed for $f^z(x)$ in which z is not an integer, but only went as far as defining $f^{-m}(x)$. Babbage, in a paper read to the Cambridge Philosophical Society on 1 May 1820 (Babbage 1822) proceeded to obtain $f^0 z = z$ from $f^n x = f^0 f^n x$ and from that to interpreting $f^{-1}x$. But these are special cases, which require special examination (like x^m itself). And in the "Philosophy of Analysis," fol. 53, he urges explicitly that the definitions of the differential calculus be such as to "include fractional indices of differentiation as well as those which are whole numbers." Not in every case can the symbol denoting the number of repeated operations be thus generalized. (Cyclical functions are a good example.) Such considerations generally belong to specific applications. Babbage's point at this juncture is that, although pure analysis is not about number, one must allow in advance for indexicals that enter definitions of operations to be limited to the integers.

108. Babbage is referring to Berkeley's criticism, in *The Analyst*, 12–16, of Newton's proof that the first derivative of x^n is nx^{n-1}. Berkeley accuses Newton of first assuming that x is increased by a finite increment o, building on this assumption in order to apply the binomial to $(x + o)^n$ and to then divide by o, only to then replace it by the contradictory assumption that, in fact, no increment in x occurred and that o is really nothing. But in that case, asserts Berkeley, one may not retain the former consequences of the rejected assumption, and is required to begin anew. Berkeley formulates the principle that Babbage, I believe, refers to as "the shifting of the hypothesis" thus: "If, with a view to demonstrate any proposition, a certain point is supposed, by virtue of which certain other points are attained; and such supposed point be itself afterwards destroyed or rejected by a contrary supposition; in that case, all the other points attained thereby, and consequent thereupon, must also be destroyed and rejected, so as from thenceforward to be no more supposed or applied in the demonstration."

109. "Philosophy of Analysis," fols. 48–50.

110. There were other problems with Lagrange's approach, of course. The very identification of the derivability of a function with its susceptibility to a Taylor expansion was being seriously questioned at the time (with Cauchy furnishing the classical counter-example of $f(x) = e^{-(1/x^2)}$, a function derivable yet nonexpandable at $x = 0$). More disturbing from Babbage's perspective, perhaps, was the fact that, regardless of the calculus, working with power series expansions required limit-based reasoning in order to decide convergence. But these were problems peculiar to Lagrange's approach itself and were not especially aggravated by Babbage's new theory of analysis.

111. "Philosophy of Analysis," fol. 51.

112. The diehard Lagrangians were Peacock and especially Herschel, who retained their early allegiances to Lagrange's version of the calculus well after its demise across the English Channel and the publication of Cauchy's algebraical epsilon-delta, limit-based definitions. Peacock explicitly deemed Cauchy's procedures an attempt "to conciliate the direct consideration of infinitesimals with the purely algebraical views of the principles of this calculus, which Lagrange first securely established . . . altogether inconsistent with the spirit and principles of symbolical algebra" (Peacock 1833, 247–48). Herschel's allegiance to Lagrange and apparent disregard of Cauchy's revision of the calculus is less explicit if equally evident in his long entry "Mathematics" in the *Edinburgh Encyclopedia*, in which Lagrange is hailed as "[a] more consummate analyst than Clairaut,—a more profound reasoner than D'Alembert,—a more comprehensive genius than Euler" (Herschel 1832, 458), and his revision of the calculus, as having accomplished "the greatest revolution which the nature of a science like mathematics could admit in the principles of its most extensive branch" (451). Cauchy is not even mentioned. In contrast, those who, like Whewell and DeMorgan, rejected Lagrange's version of the calculus during the 1830s did so for reasons precisely opposite to those of Babbage. For them the differential calculus was not an abstract system *applied* to limits, but a systematic articulation *of* the very notion of limit.

113. "Philosophy of Analysis," fol. 41.

114. "Philosophy of Analysis," fol. 53.

115. "Philosophy of Analysis," fol. 54.

CHAPTER SIX

1. Lambert 2013. Lambert's paper is the most significant addition to the literature related to Peacock in recent years, and adds an important dimension to the account of the making of the *Treatise* offered in my earlier work. The fact that I remain in some disagreement with his assessment of my reading of the *Treatise* in no way detracts from the importance of his work.

2. Enros 1979, 120, citing Herschel's obituary of Peacock (Herschel 1858, 536).

3. "The true faith" was a catchword employed by the young dissenters to denote Lagrange's approach to the calculus. See, for example, Babbage to Herschel, 13 June 1813, RS HS 2.15; Babbage to Herschel, 8 December 1815, RS HS 2.50; Herschel to Babbage, 14 July 1816, RS HS 2.64; Babbage to Herschel, 20 July 1816, RS HS 2.65; Peacock to Herschel, 3 December 1816, RS HS 13.247; and Whewell to Herschel, 6 March 1817, cited in Todhunter 1876, ii,16.

4. He not only spearheaded the reform process at Cambridge, but also drew most of the fire. The university granted the young examiners of each year extensive freedom in framing the questions on the Tripos. This afforded Peacock, as moderator in 1817 and 1819, the opportunity to introduce the differential notation in his share of the questions, for which he encountered considerable opposition. Writing to Herschel in December 1816, Peacock was confidently optimistic: "We are proposing college reforms, which will introduce the true faith in Trinity; at least if our wise seniority will adopt them. . . . Trust me, the golden age of the University is approaching." (Peacock to Herschel, 3 December 1816, RS HS 13.247.). His optimism proved premature. "White and Fallows [proved] entirely of the old school," he wrote a few weeks later, "& the influence of their examination was so great as completely to overpower my examination. The introduction of d's into the papers excited much remorse . . . & I believe that I may consider myself as owing entirely to the success of the Johnians in the examination for my escape from some public proceedings against me" (Peacock to Herschel, undated, postmark: 1817, RS HS 13.248). Peacock's initiative was publicly criticized by D. M. Peacock (1819). But the opposition from without eventually died down. For further details see Dubbey 1978, chap. 3.

5. Nor can I detect in this phase of Peacock's development the same kind of ambivalating impact that deliberating how best to teach mathematics would have a little later on Whewell.

6. Even Verburgt, whose work on the second generation of Cambridge mathematical reformers comes closest to the account presented here, in attempting to steer a middle course between viewing the development of modern algebra in Britain in traditional revolutionary and evolutionary terms, ends up offering an essentially gradualist, if second-order account of the emergence of the notion of abstraction itself. See especially Verburgt 2016, 33.

7. Lacroix 1816, iv.

8. *Passages* was published in 1864, while MS 13, though many years in the making, contains two explicit references in Babbage's hand to 1869.

9. Fisch 2014.

10. Lacroix 1816, 465–579, at 539–40.

11. Herschel wrote a considerable amount of mathematics during the mid-1810s (for a detailed survey, see Enros 1979, chap. 5), but refrained from making general statements about the nature of mathematics in general or any of its branches. Still, a paper he read at the Royal Society on 14 December 1815 (Herschel 1816) suggests that he had adopted Babbage's prohibition against the employment of infinite series in pure analysis. This may explain his apparent lack of enthusiasm for Lagrange in his contribution to Lacroix. Yet, in a textbook aimed at dislodging the university's century-long commitment to Newtonian fluxions, Herschel's agnostic rhetoric might well have been a contrived tactic.

12. Lacroix 1816, 612, Note B. Compare Herschel and Babbage's joint preface to the *Memoirs of the Analytical Society*, which focuses exclusively, not on the conceptual merits of the Lagrangian calculus, but on the "accurate simplicity of its language," "the symmetry," and "the conciseness of its notation" (Herschel and Babbage 1813, i–ii)!

13. Lacroix 1816, 614.

14. Lacroix 1816, 618. His reference to "the consideration of motion" may be hinting at Maclaurin's (1742) important elaboration of fluxional claculus.

15. Lacroix 1816, 596.

16. Lacroix 1816, 619.

17. Peacock to Herschel, 14 November 1816, RS HS 13.246.

18. Referring presumably to Waring 1770.

19. Peacock to Herschel, 12 December 1816, RS HS 13.247.

20. See especially Richards 1991, 2007, and 2011a, as well as Durand-Richard 1990.

21. For Locke's comparison of mathematical and religious knowledge, see Locke 1836, book 4, chap. 10, entitled "Of Our Knowledge of the Existence of a God," and sections 7–8 of chap. 12.

22. Locke 1836, book 2, chap. 1, quote at 51.

23. These, however, were based on the sharp and sobering account in book 1 of the *Novum Organon* (Aph. 38 and following) of the four idols, or categories of impediments to the acquisition and retention of true knowledge, for which "the formation of ideas and axioms by true induction is no doubt the proper remedy to be applied" (Aph. 40). This part of Bacon's project certainly owes to keen reflection rather than sensation, though largely unconcerned with epistemology or the nature of truth.

24. Other work on Herschel and Babbage reveals that other, non-Lockean cultural factors were also involved, in addition to those pertaining to their mathematics proper. For an interesting attempt to view Herschel's and especially Babbage's early approach to algebraic analysis as inspired by their fascination with contemporary industrial culture, see Schaffer 1994 and 1996; and Ashworth 1994 and especially 1996.

25. Whewell 1819 and 1823. During the 1830s he published an additional four such texts (1832a, 1832b, 1833b, and 1837a), as well as a new edition of his *Treatise on Dynamics* in two parts (1832c and 1834a).

26. The problem was to explain to the uninitiated what motivates casting a realist picture of reality in the terms of a purely formal, nonrealist mathematical system—how to explain identifying real physical velocities and accelerations with coefficients of a power series without compromising realism? Evidence of Whewell's growing awareness of the threat of such instrumentalist moves one finds first in his abandonment and eventual all-out denunciation of the use of virtual velocities and other variational techniques

in dynamics, and the conflation of statics and dynamics by virtue of such techniques in standard French texts. For further details see Fisch 1991a, 31–46.

27. I take this opportunity to thank Joan Richards for sharing with me her work in progress on the differences between the two universities. During the period in question Oxford refused to enroll dissenters. Cambridge, in contrast, accepted them but would not grant them degrees. (Or as Richards likes to put it: at Oxford they weren't allowed in, while at Cambridge they weren't allowed out.) As a result, while the Oxford curriculum could take the principles of the faith for granted, Cambridge could not. They had to be taught and defended. This, she suggests, explains why logic was stressed at Oxford and mathematics at Cambridge.

28. Herschel and Babbage both sought and acquired Cambridge positions in the course of their careers, but to no real effect. After graduating with the highest honors in January 1813, Herschel left Cambridge, became a fellow of St. John's in March, spent the fall at Cambridge, but left in January 1814, initially to study law. He came back briefly to tutor the following year, but apparently hated it, left, and never returned. Desperate for a position after his marriage in July 1814, Babbage applied unsuccessfully for professorships both at Cambridge and Edinburgh. He finally obtained the prestigious Lucasian Professorship at Cambridge (1828–39), but seems never to have lectured or to have taken much interest again in the university's mathematics curriculum.

29. Their description as Lockeans owes to Lambert 2013, who takes it from Richards.

30. Maseres 1758 and 1800; Frend 1796. For a discussion of their proposals, the manner in which they wreaked havoc in the theory of algebraic equations, and Peacock's later reaction to them, see Pycior 1981, 27–31.

31. Peacock 1833, 190–91. On Herschel's very different treatment of the problem of negatives and imaginaries, see Herschel 1833, 439 and 446; and Fisch 1994c, n. 58.

32. See Peacock 1830, 110–13, where he concludes that, as opposed to geometry, in a "science of general symbols, which by themselves express no relations or magnitudes . . . there are, properly speaking, no axioms, since the propositions, *immediately* deducible from the definitions and assumptions, must be considered rather as the necessary and immediate consequences of defined operations, than the necessary and self-evident results of reasoning" (112–13).

33. Peacock 1830, 110–13.

34. A point well taken by Verburgt (2016).

35. Cited above in chapter 5, note 80 and associated text.

36. Peacock to Babbage, 7 May 1822, British Museum, Add. MS 37182, No. 323; see Lambert 2013, text to n. 22.

37. For a detailed account of the formation of the Cambridge Philosophical Society, its early membership, and activities, see President John Willis Clark's 1890 Address in vol. 7 of the society's *Proceedings*.

38. The three papers published in the *Transactions* were Babbage 1822, read by the author on 1 May 1820, and Babbage 1827b and 1827a, both read by Peacock on 10 December 1821 and 3 May 1824, respectively. A fourth, entitled "On the Principles of Mathematical Notation" (read by Peacock on 12 December 1825), was never published.

39. See DeMorgan 1847, 91. DeMorgan also claims that Peacock's entry "was separately published, in parts, in eighteen-twenty-five or twenty-six." There is no record of

this. DeMorgan may have wrongly assumed that the papers Peacock presented at society meetings in 1826 and 1827 were also published at the time.

40. As I remarked in chapter 5, the empirical and inductive nature of Peacock's arithmetic project not only merits his inclusion among the inner circle of Snyder's (2011) Baconian "Breakfast Club," but would have added to her story a mathematical dimension it seriously lacks. In one of the best-crafted chapters of her book, entitled "Mapping the World," she sets side by side three of the great mapping projects of the 1830s—Herschel's astronomical catalogs, Whewell's worldwide study of the tides, and Jones's nationwide tithe maps. Snyder seemingly focuses on mappings of the natural world—the land, the sea, and the stars. But Jones's mapping focused not merely on the natural terrain, but on the results of human parcelation and ownership. Peacock's arithmetic project would have provided Snyder with a mapping, comparable in scope and detail, that focused on an even more conspicuously universal element of human nature, life, and culture. Yet unlike the purely synchronic mappings of the other three, Peacock's arithmetic also contained a distinctly *diachronic*, historical dimension that was designed to facilitate an account of the *emergence* of the arithmetical present. Had it been included in Snyder's chapter, it might have paved the way for the inclusion of another great survey undertaken by a former Breakfast Club member conspicuously absent from her account—Whewell's unprecedented 1837 mapping of the history of the inductive sciences!

41. Durand-Richard 2010; Lambert 2013.

42. See also Durand-Richard 1990.

43. Lambert 2013, 281.

44. Lambert 2013, 282.

45. Lambert 2013, 282.

46. Peacock 1830, v.

47. As we shall see in the following chapter, this is how Whewell, in 1833, in an incredible combination of haste and wishful thinking, initially interpreted the *Treatise*, in order to assimilate it to his own developing system of thought.

48. Peacock 1830, xi.

49. Peacock 1834, 194. There are no substantial differences between the theories of algebra presented in Peacock 1830 and 1834. The latter, however, not being a textbook, dwells at greater length and in greater detail on the meta-mathematics at stake, and is, therefore, a source better suited for the concerns of the present study. In what follows, page references to this work will be incorporated in the text. Unless indicated otherwise, all italics are Peacock's.

50. Peacock further emphasizes the point in a footnote attached to the word *generalization* in the passage cited, concluding that: "The process, therefore, by which we pass from one science to the other is not an ascent from particulars to generals, which is properly called *generalization*, but one which is essentially arbitrary."

51. Of course, that suggestion *per se* need not entail inclusion. $a^m \cdot a^n = a^{m+n}$, for example, certainly suggests $(d^n/dx^n)(d^m u/dx^m) = d^{n+m}u/dx^{n+m}$, though neither formula includes the other.

52. See also Peacock 1830, xv–xvii.

53. Which is indeed the term adopted by Peacock in the second edition, fifteen years later. Peacock's change of heart in this respect is taken up in the last section of chapter 7.

54. On the difference between Frend's and Peacock's understanding of "the Lockean imperative," as it were, see Lambert 2013, 287–89.

55. Which, as noted in the previous chapter, renders groundless Dubbey's (1978) and Becher's (1980b) contention that Peacock's symbolical algebra resembled that of Babbage to the point of plagiarism.

56. Or as Gregory (1840, 5) would later propose more generally: $f_m(a) \cdot f_n(a) = f_{m+n}$ and $f_m f_n(a) = f_{mn}$.

57. Babbage had aimed even higher than that; he believed his was an attempt to discern the very grammar of human thought.

58. See Berkeley 1734, §XIII–XVI.

59. Peacock 18343, 204–23.

CHAPTER SEVEN

1. To the extent that, as Hankins (1980, 249) notes, "from the 1820s through the 1840s the study of the foundations of algebra became almost a British monopoly."

2. As will become apparent, both Hamilton in his later work and Gregory in the little he wrote on the foundations of algebra a little earlier present their positions as being in fundamental agreement with Peacock's. Hamilton's change of heart, I shall argue, was in all probability due to Peacock's own change of heart in the second edition of the *Treatise*. Gregory's supposed agreement—formulated several years before the publication of Peacock's second edition—rested, I shall suggest, on a misunderstanding of the inherently hybrid nature of Peacock's initial position.

3. Although, as we shall see, Hamilton, in later years thought he did.

4. See especially Verburgt 2016, 2015, and forthcoming article on Gregory and Ellis.

5. I find Verburgt's distinction between the symbolical algebras of Peacock and Gregory and the abstract algebra of Hamilton's quaternions extremely useful, as long as one is careful not to classify Peacock (as did Gregory) as a symbolical algebraist *simpliciter*—as Verburgt on occasion seems to do. Peacock's algebra was more profoundly hybridic and unstable than anything published in its wake. No one else came remotely near to viewing algebra as actually comprising *two* incongruent algebras, as he did.

6. DeMorgan's review of the *Treatise* was published in the *Quarterly Journal of Education* in 1835. On the development of DeMorgan's view of algebra, see the seminal Richards 1987.

7. DeMorgan 1840, quoted and discussed in Richards 1987, 15–17 (italics added). "This formulation," she aptly notes, "reverses the order Peacock's principle established in algebra. Rather than moving [by arithmetical suggestion] from meaning to abstract forms, the development is from abstract forms to meaning" (16).

8. DeMorgan 1842a.

9. DeMorgan 1837, 174.

10. DeMorgan 1842a, 288–89.

11. I am referring especially to Richards's (1991) insightful comments, first, on DeMorgan's early definitions of mathematics as "the consideration of the properties of space, number and magnitude in general" (DeMorgan's speech at the opening of his mathematical classes at the newly founded University of London, cited in Richards 1991,

308), and especially his insistent grounding of the calculus on a detailed articulation of the concept of limit, in explicit contrast to the society's Lagrangianism, in the lengthy introduction to DeMorgan 1842b—a work originally published in installments, in successive numbers of the Library of Useful Knowledge, beginning in 1836.

12. Richards 2011b nicely captures what she terms the "Janus-faced vision" of DeMorgan's mathematics, which she interprets, as in her earlier work, as owing to his essentially historical view of mathematics.

13. Hamilton 1837, 295.

14. Inserted as an afterthought.

15. Hamilton to Peacock, 23 July 1835, Trinity College Library, Add.ms.b.49[44]. In writing to others, Hamilton was far more blunt. In 1831, soon after reading the *Treatise*, he wrote to Aubrey De Vere: "I differ from my great contemporaries, my 'brother band,' not in transient or accidental, but in essential and permanent things: in the whole spirit and view with which I study Science" (Hamilton to De Vere, 9 February 1831, cited in Graves 1882, 1:519, and Hankins 1980, 249.)

16. Graves 1885, 2: chap. 18; Hankins 1980, chap. 6.

17. Hamilton's "On Quaternions: Or a New System of Imaginaries in Algebra" was published in eighteen installments in the *Philosophical Magazine* between 1844 and 1850, and his "On Symbolical Geometry," in ten installments in the *Cambridge and Dublin Mathematical Journal* between 1846 and 1849.

18. Hamilton's debt to Coleridge in this respect mainly concerns the latter's much-labored distinction (inspired by Kant) between Reason and Understanding. In Coleridge's opinion, the former alone constitutes the "scientific faculty" in being "the power by which we become possessed of Principle . . . and of Ideas" prior to and independently of the senses (*The Friend*, No. 7, 28 September 1809, in Coleridge 1969, 4: part 2, p. 104). See also Coleridge 1861, 167–204. Interestingly, though, Coleridge stresses the similarity between "Reason of contemplation" (as opposed to understanding) and sensual perception. "Reason" he writes, "is much nearer to Sense than to Understanding: For Reason . . . is a direct aspect of truth, an inward beholding, having a similar relation to the intelligible or spiritual, as Sense has to the material or phenomenal" (1861, 174). The science of mathematics, obtained, according to Coleridge, by the contemplation of "pure sense" (183), pertains therefore quite explicitly to a notion of truth by correspondence in the realm of ideas akin to its experiential counterpart in the realm of fact. For Hamilton's debt to Coleridge, see Hankins 1980, 255–57; and Bloor 1981, 209–20.

19. Which, as Hankins (1977, 176) shows, we now know is impossible.

20. Hamilton 1853, 1.

21. Hamilton 1853, 14–16.

22. Hamilton to Peacock, 13 October 1846 (from a draft); Graves 1885, 2: 527–29, partly cited in Hankins 1980, 250.

23. Hamilton to Peacock, 13 October 1846.

24. The paper was presented to the Royal Society of Edinburgh in 1838 and published in its transactions in 1840.

25. Gregory 1840, 1.

26. Verburgt (2015) distinguishes interestingly between abstract and symbolical

algebras. He deems the former modern and the latter "pre-modern" (33). There is merit to the distinction, especially when assessing their works retrospectively from a modern perspective. But when studied with the prospective, forward-looking gaze of the actors under study, the differences between Peacock's symbolical algebra and Gregory's further abstraction are profound.

27. Gregory 1840, 1; emphasis added.

28. See, for example, Close 2010, especially chap. 3.

29. Trinity College Library, Whewell Papers R.18.178.

30. Whewell 1840, 1: book 2, chaps. 3 and 8, respectively. A rare adoption of Kantian terminology on Whewell's part, who, though deeply endebted to Kant, carefully avoided Kantian jargon.

31. Later, during the late 1820s and 1830s, Lagrange's approach was gradually adopted, less in mechanics proper than in wave-optics and the theory of heat. For a detailed study of the "increasing preference for the new works of Fresnel and Fourier over those of Laplace and his disciples," see Smith and Wise 1989, 155–68.

32. A tendency well evidenced in Peacock 1833, 186–88, and most conspicuously in Herchel 1830 and 1833.

33. Smith and Wise 1989, 155–68.

34. For an initial account of the problematic implications of such a position, see Fisch 1991a, 33–35, summarized briefly in chapter 6, note 26, above.

35. Fisch 1991, chaps. 2–3.

36. Contrary to Herschel's influential work of 1830, Whewell was adamant at the time to formulate a philosophy of science capable of accommodating physics at its fully mathematicized best. Herschel, as noted, kept his philosophical musings on empirical science and mathematics quite separate, limiting his philosophy of science, in true Baconian fashion, to the observational and experimental aims and methods of natural philosophy, thus evading all discussion of mathematical physics proper.

37. See note 30, above. Although he had studied Kant closely during the second half of the 1820s and had taken extensive notes on the First Critique, in which, as Ducheyne (2011) has meticulously recorded, Kant's terminology figures prominently, Whewell's published and unpublished work (by which I mean draft essays and chapters of his own, as opposed to his reading notes), carefully avoids any mention of Kant's distinct terminology. This is all the more puzzling if one accepts Ducheyne's (2011) main claim that Whewell's philosophy "took Kant's philosophy as a point of departure for his doctrine of Fundamental Ideas, was unsatisfied with Kant's philosophy of science . . . and that he correspondingly tried to modify the Kantian project" (38). I fully agree that in the course of his philosophical journey, Whewell endorsed major components of the transcendental philosophy of the "metaphysical genius in Germany," as Kant is referred to in the *Philosophy of the Inductive Sciences*—very much as Ducheyne describes. But I disagree with Ducheyne that the problem with which he was grappling resembled the problem Kant had awoken to late in life in his *Opus Postumum*, as described by Friedman 2003 and Ducheyne. Whewell's meta-scientific deliberations began long before he encountered Kant. After doing so, the problem he faced was not to bridge metaphysics and the empirical sciences, but to somehow account for science's *realism*, without relinquishing his

Kantian doctrine of Fundamental Ideas, to somehow bridge, if you wish, the noumenal-phenomenal divide—something Kant never considered. In this sense, despite its rare philosophical sophistication, his hybrid theory of science remained hopelessly split.

38. In dealing with an original thinker whose views took form in the course of grappling with foundational issues, one cannot hope to properly understand his works merely by reading them, as R. G. Collingwood liked to say (Collingwood 1939, chap. 5, cited above, chapter 1, note 8). They need to be read as what their authors considered to be viable solutions to the problems they believed they were facing. But as argued in chapter 4, their published works seldom make that explicit. They make little if any mention of the false starts, the dead ends, the aborted attempts, or the difficulties they ran into en route to their final formulation. Their unpublished papers, if they bothered at all to save them (as did Whewell, but not Peacock), are usually also of little help, consisting for the most part of discarded drafts of texts written to be published in the same highfaluting, confident style.

But by carefully comparing an author's published and unpublished work it is sometimes possible to glimpse the deliberative process that resulted in publication. The trick is to read their notebooks and drafts with an eye, less for continuations, i.e. for *anticipations* of their final positions, than for *dis*continuations, i.e. for what they elected *not* to publish. For a text deemed unfit for publication, is a text they obviously considered either superfluous, unsuitable or problematic. By working back from the contents of an author's wastepaper basket, so to speak, rather than from what he or she decided to publish, one can sometimes learn a lot about the straws they were grasping at and the options they considered viable and eventually came to deem inappropriate. In this sense, Whewell's writing on algebra is a case in point.

39. "Philosophy of the Pure Sciences," 105.

40. Five years later, Whewell (1838) presented a book-length textbook exposition of "The Doctrine of Limits."

41. "Philosophy of the Pure Sciences," 92 and following.

42. "Philosophy of the Pure Sciences," 82–83 (italics added).

43. On Whewell's notion of necessity, see Fisch 1985, 1987, 1991a, 154–63, and 1991c.

44. "Philosophy of the Pure Sciences," 83–84 (italics added).

45. "Philosophy of the Pure Sciences," 87.

46. The idea of a mathematical operation or operator might have proved to be a better candidate, but he seems not to have considered it.

47. "Philosophy of the Pure Sciences," 86.

48. For the two earliest expressions of his notion of conceptual truth, see Whewell 1835 for geometry and the calculus; and Whewell 1834b for the axioms of causation. To the best of my knowledge, the term "fundamental idea" occurs for the first time in the former (17).

49. This, however, leads to other problems: a^n is perhaps too easy a case since it involves no obvious exceptions, but what of m/n nor $\log(n)$? Is one allowed on the basis of the principle of absolute generality to substitute 0 for n? A decade later DeMorgan (1844) would raise this problem forcefully with respect to divergent series, and answer it with a forceful yes! See Richards 1987, 26–29; and 2011b.

50. "Philosophy of the Pure Sciences," 85 (italics added).

51. "Philosophy of the Pure Sciences," 86.

52. Describing "such merely notional and hypothetical '*virtual* velocities' [which] have no connection with physical and *actual* velocities produced by forces" as "*geometrical phantoms*" (Whewell 1832a,viii–ix). See also chapter 6, note 26, above; and Fisch 1991a, 45 and additional references there.

53. "Philosophy of the Pure Sciences," 86.

54. First to be fully clarified in a paper written that year on the nature of the truth of the laws of motion (Whewell 1834b) that was read to the Cambridge Philosophical Society on 17 February 1834 and published in the *Transactions* a year later.

55. "Philosophy of the Pure Sciences," 87 (italics added).

56. Whewell 1840, 1:142–43.

57. Hankins 1980, 254.

58. Fisch 1991a, 140–42.

59. Peacock 1842, vi–vii.

60. Peacock 1842, vi–vii.

61. Peacock 1845, 59–60.

62. Peacock 1842, vii.

63. The inconsistent capitalizing is Peacock's. Arithmetical and symbolical algebra, geometry, and so forth, spelled with lower case throughout volume 1, are all capitalized, Carlyle-style, in volume 2.

64. Crowe 1994, 14, 15, 34, 108.

65. Peacock 1845, iv.

66. Peacock 1842, iii.

REFERENCES

PRIMARY SOURCES

Unpublished Works

Babbage, Charles. "The History of the Origin and Progress of the Calculus of Functions during the Years 1809, 1810 . . . 1817." Oxford, History of Science Museum, MS Buxton 13. Cited as MS 13.

———."Essays on the Philosophy of Analysis." c. 1821. London, British Museum, Add. MS 37202.

Herschel, J. F. W. *Letters and Papers of Sir John Herschel From the Archives of the Royal Society*. Ed. Paul L. Kesaris. Microform (28 microfilm reels). Frederick, MD: University Publications of America, c. 1990. References follow the notation adopted in Michael Crowe's *A Guide to the Manuscripts and Microfilms*: RS HS reel #.document #).

Whewell, W. "The Philosophy of the Pure Sciences." 1833. Cambridge, Trinity College Library, Whewell Papers R.18. 178.

Published Works

Agnesi, Maria G. A. M. 1801. *Analytical Institutions, in Four Books: Originally Written in Italian, by Donna Maria Gaetana Agnesi, Professor of the Mathematics and Philosophy in the University of Bologna . . .* Trans. John Colson; rev. John Hellins. London: Taylor and Wilks.

Babbage, Charles. 1813. "On Continued Products." In Herschel and Babbage 1813, 1–31.

———. 1815. "An Essay towards the Calculus of Functions, Part I." *Philosophical Transactions* 105: 389–423.

———. 1816. "An Essay towards the Calculus of Functions, Part II." *Philosophical Transactions* 106: 179–256.

———. 1817. "Observations on the Analogy Which Subsists between the Calculus of Functions and Other Branches of Analysis." *Philosophical Transactions* 107: 197–216.

———. 1820. *Examples of the Solutions of Functional Equations*. Cambridge: Deighton.

———. 1822. "Observations on the Notation Employed in the Calculus of Functions." *Transactions of the Cambridge Philosophical Society* 1: 63–76.

———. 1827a. "On the Determination of the General Term of a New Class of Infinite Series." *Transactions of the Cambridge Philosophical Society* 2: 217–25.

———. 1827b. "On the Influence of Signs in Mathematical Reasoning." *Transactions of the Cambridge Philosophical Society* 2: 325–77.

———. 1837. *The Ninth Bridgewater Treatise—A Fragment*. London: John Murray.

———. 1864. *Passages from the Life of a Philosopher*. London: Longman. Reprinted with new introduction by Martin Campbell-Kelly. New Brunswick, NJ: Rutgers University Press, 1994. Pagination according to reprint.

Berkeley, George. 1734. *The Analyst: Or, a Discourse Addressed to an Infidel Mathematician, Wherein It Is Examined Whether the Object, Principles, and Inferences of the Modern Analysis Are More Distinctly Conceived, or More Evidently Deduced, Than Religious Mysteries and Points of Faith*. London: J. Tonson in the Strand.

Buxton, Harry W. 1988. *Memoir of the Life and Labours of the Late Charles Babbage Esq. F.R.S.* Ed. Anthony Hyman. Cambridge, MA: MIT Press.

Coleridge, Samuel T. 1861. *Aids to Reflection*. 9th ed. London: Moxon.

———. 1969. *The Collected Works of Samuel Taylor Coleridge*. London: Routledge and Kegan Paul.

DeMorgan, Augustus. 1835. Review of Peacock 1830. *Quarterly Journal of Education* 17: 91–110; 18: 293–311.

———. 1836. *A Treatise on the Calculus of Functions (Extracted from the Encyclopaedia Metropolitana)*. London: Baldwin and Cradock.

———. 1837. "On the Foundations of Algebra, No. I." *Transactions of the Cambridge Philosophical Society* 7: 173–87.

———. 1840. "Negative and Impossible Quantities." *Penny Cyclopedia*, suppl. vol., 133–34.

———. 1842a. "On the Foundations of Algebra, No. II." *Transactions of the Cambridge Philosophical Society* 7: 287–300.

———. 1842b. *The Differential and Integral Calculus*. London: Baldwin and Cradock.

Ditton, Humphry. 1706. *An Institution of Fluxions: Containing the First Principles, the Operations, with Some of the Uses and Applications of That Admirable Method; According to the Scheme Perfix'd to His Tract of Quadratures, by (its First Inventor) the Incomparable Sir Isaac Newton*. London: W. Botham for J. Knapton.

Frend, William. 1796. *The Principles of Algebra*. London: J. Davis.

Graves, Robert P. 1882, 1885, 1889. *Life of Sir William Rowan Hamilton*. 3 vols. Dublin: Hodges, Figgis and C.; London: Longmans.

Gregory, Duncan F. 1840. "On the Real Nature of Symbolical Algebra." *Transactions of the Royal Society of Edinburgh* 14: 208–16. Reprinted in Gregory 1865, 1–13. Pagination according to reprint.

———. 1865. *The Mathematical Writings of Duncan Farquharson Gregory Late Fellow of Trinity College Cambridge*. Ed. William Walton, with Biographical Memoir by Robert Leslie Ellis. Cambridge: Deighton.

Hamilton, William Rowan. 1837. "Theory of Conjugate Functions, or Algebraic Couples;

With a Preliminary And Elementary Essay on Algebra as the Science of Pure Time." *Transactions of the Royal Irish Academy* 17 (1): 293–422.

———. 1853. *Lectures on Quaternions*. Dublin: Hodges and Smith.

Herschel, John F. W. 1813a. "On Trigonometrical Series, Particularly Those Whose Terms Are Multiplied by the Tangents, Cotangents, Secants, &c. of Quantities in Arithmetic Progression, Together with Some Singular Transformations; with Notes Relating to a Variety of Subjects Connected with the Preceding Memoir." In Herschel and Babbage 1813, 33–64.

———. 1813b. "On Equations of Differences and Their Application to the Determination of Functions from Given Conditions." In Herschel and Babbage 1813, 65–114.

———. 1813c. "On a Remarkable Application of Cotes's Theorem." *Philosophical Transactions* 103, 8–26.

———. 1814. "Consideration of Various Points of Analysis." *Philosophical Transactions* 104: 440–68.

———. 1819a. "Preface to the Essays." In Spence 1819, xxvii–xxxii.

———. 1819b. "Note on Essay II." In Spence 1819, 151–70.

———. 1820. *A Collection of Examples of the Applications of the Calculus of Finite Differences*. Cambridge: Deighton.

———. 1830. *A Preliminary Discourse on the Study of Natural Philosophy. The Cabinet Cyclopaedia, Natural Philosophy*. London: Longmans.

———. 1832. "Mathematics." In *Edinburgh Encyclopedia*, ed. David Brewster, vol. 12, 434–59.

———. 1833. "Mechanism of the Heavens." Review of Sommerville 1832, and Laplace's *Exposition du système du monde* of 1808. *Quarterly Review* 99. Reprinted in Herschel 1857, 21–62.

———. 1857. *Essays from the Edinburgh and Quarterly Reviews with Addresses and Other Pieces*. London: n.p.

———. 1858. "[Obituary Notice of] The Late Very Rev. Dr. Peacock, Dean of Ely." *Royal Society Proceedings* 9 (1858): 536–43.

Herschel, John F. W., and Charles Babbage. 1813. *Memoirs of the Analytical Society, 1813*. Cambridge: n.p.

Jones, Richard. 1859. *Literary Remains Consisting of Lectures and Tracts on Political Economy, by the Late Rev. Richard Jones*. Ed. William Whewell. London: John Murray.

Lacroix, Silvestre F. 1797–1800. *Traité du calcul différentiel et du calcul intégral*. 3 vols. Paris: Chez Courcier.

———. 1802. *Traité élémentaire de Calcul différentiel et de Calcul intégral*. Paris: Duprat; later editions published by Courcier.

———. 1816. *An Elementary Treatise on the Differential and Integral Calculas*. English translation of parts 1 and 2 of Lacroix 1802 by Charles Babbage, George Peacock, and William Herschel, with notes by Peacock and Herschel and an appendix on "differences and series" by Herschel. Cambridge: Deighton.

Lagrange, Joseph-Louis. 1797. *Théorie des fonctions analytiques contenant les principes du calcul différentiel dégagés de toute considération d'infiniment petits, d'évanouissans, de limites et de fluxions, et réduits à l'analyse algébrique des quantités finies*. Paris: Imprimerie de la République.

Laplace, Pierre-Simon (Marquis de Laplace.) 1798–1825. *Traité de mécanique céleste.* 5 vols. Paris.

Locke, John. 1836. *An Essay Concerning Human Understanding.* 27th ed. London: Tegg and Son.

Maclaurin, Colin. 1742. *A Treatise on Fluxions, in Two Books.* Edinburgh: T. W. and T. Ruddimans.

Maseres, Francis. 1758. *A Dissertation on the Use of the Negative Sign in Algebra.* London: J. Davis.

———. 1800. *Tracts on the Resolution of Affected Algebraick Equations by Dr. Halleys, Mr. Raphsons, and Sir Isaac Newton's Methods of Approximation.* London: J. Davis.

Peacock, Daniel M. 1819. *A Comparative View of the Principles of the Fluxional and Differential Calculus Addressed to the University of Cambridge.* Cambridge: Deighton.

Peacock, George. 1820. *A Collection of Examples of the Applications of the Differential and Integral Calculus.* Cambridge: Deighton.

———. 1829. "Arithmetic." In *Encyclopaedia Metropolitana,* vol. 1, 372–523. London: B. Fellowes et al.

———. 1830. *A Treatise on Algebra.* Cambridge: Deighton.

———. 1834. "Report on the Recent Progress and Present State of Certain Branches of Analysis." In *Report on the Third Meeting of the British Association for the Advancement of Science,* 185–352. London: n.p.

———. 1842. *A Treatise on Algebra. Vol. 1: Arithmetical Algebra.* Cambridge: Deighton.

———. 1845. *A Treatise on Algebra. Vol. 2: On Symbolical Algebra and Its Applications to the Geometry of Position.* Cambridge: Deighton.

Playfair, John. 1808. "Review of Laplace, *Mécanique céleste.*" *Edinburgh Review* 11: 243–84. Reprinted in Playfair 1822, vol. 4, 261–330. Pagination according to reprint.

———. 1822. *The Works of John Playfair, Esq.* 4 vols. Edinburgh: Archibald Constable and Co.

Simpson, Thomas. 1805. *The Doctrine and Application of Fluxions: Containing (Besides What is Common on the Subject) a Number of New Improvements in the Theory, and the Solutions of a Variety of New and Very Interesting Problems in Different Branches of the Mathematics.* London: H. D. Symonds, Knight, and Compton.

Sommerville, Mary. 1832. *Mechanism of the Heavens.* London: n.p.

Spence, William. 1819. *Mathematical Essays by the Late William Spence.* Ed. J. F. W. Herschel. Edinburgh: Oliver and Boyd.

Todhunter, Isaac. 1876. *William Whewell D.D.—An Account of His Writings, with Selections from His Literary and Scientific Correspondence.* 2 vols. London: Macmillan and Co.

Ward, John. 1707. *The Young Mathematician's Guide—Being a Plain and Easy Introduction to the Mathematicks in Five Parts . . .* London: Printed by E. Midwinter, for John Taylor.

Whewell, William. 1819. *An Elementary Treatise on Mechanics.* Cambridge: Deighton.

———. 1823. *A Treatise on Dynamics.* Cambridge: Deighton.

———. 1832a. *The First Principles of Mechanics with Historical and Practical Illustrations.* Cambridge: Deighton.

———. 1832b. *An Introduction to Dynamics Containing the Laws of Motion and the First Three Sections of the Principia*. Cambridge: Deighton.

———. 1832c. *On the Free Motion of Points and On Universal Gravitation, Including the Principal Propositions of Books I and III of the Principia: The First Part of a New Edition of A Treatise on Dynamics*. Cambridge: Deighton.

———. 1833. *Analytical Statics*. Cambridge: Deighton.

———. 1834a. *On the Motion of Points Constrained and Resisted, and the Motion of Rigid Bodies: The Second Part of a New Edition of A Treatise on Dynamics*. Cambridge: Deighton.

———. 1834b. "On the Nature of the Truth of the Laws of Motion." *Transactions of the Cambridge Philosophical Society* 5: 149–72.

———. 1835. *Thoughts on the Study of Mathematics as Part of a Liberal Education*. Cambridge: Deighton.

———. 1837a. *Mechanical Euclid*. Cambridge: Deighton.

———. 1837b. *History of the Inductive Sciences from the Earliest to the Present Time*. 3 vols. London: John W. Parker.

———. 1838. *The Doctrine of Limits with Its Applications; Namely, Conic Sections, the First Three Sections of Newton, the Differential Calculus*. Cambridge: Deighton.

———. 1840. *The Philosophy of the Inductive Sciences Founded upon Their History*. 2 vols. London: John W. Parker.

———. 1844. "On the Fundamental Antithesis of Philosophy." *Transactions of the Cambridge Philosophical Society* 7 (2): 170–81.

———. 1847. *The Philosophy of the Inductive Sciences Founded upon Their History*. 2 vols. 2d ed. London: John W. Parker.

———. 1859. "Prefatory Notice." In Jones 1859, ix–xl.

Wood, James. 1806. *The Elements of Algebra: Designed for the Use of Students in the University*. Cambridge: At the University Press.

Woodhouse, Robert. 1803. *The Principles of Analytical Calculation*. Cambridge: Deighton.

———. 1809. *A Treatise on Plane and Spherical Trigonometry*. Cambridge: Deighton.

———. 1812. *An Elementary Treatise on Astronomy*. Cambridge: Deighton.

Secondary Sources

Aggasi, Joseph. 1964. "On the Nature of Scientific Problems and their Roots in Metaphysics." In Bunge 1964, 189–211.

Altham, James E. J., and Ross Harrison (eds.). 1995. *World, Mind and Ethics: Essays on the Ethical Philosophy of Bernard Williams*. Cambridge: Cambridge University Press.

Anderson, Joel. 1996. "The Personal Lives of Strong Evaluators: Identity, Pluralism, and Ontology in Charles Taylor's Value Theory." *Constellations: An International Journal of Critical and Democratic Theory* 3 (1): 17–38.

Anstey, Peter (ed.). 2013. *The Oxford Handbook of British Philosophy in the Seventeenth Century*. Oxford: Oxford University Press.

Ashworth, William J. 1994. "The Calculating Eye: Baily, Herschel, Babbage and the Business of Astronomy." *British Journal for the History of Science* 27 (4): 401–44.

———. 1996. "Memory, Efficiency and Symbolic Analysis: Charles Babbage, John Herschel and the Industrial Mind." *Isis* 87: 629–53.

———. 2008. "The Analytical Society 1812–13." In *Oxford Dictionary of National Bibliography*. Oxford: Oxford University Press.

Bartley, William W., III. 1984. *The Retreat to Commitment*. LaSalle, IL: Open Court.

Becher, Harvey W. 1980a. "William Whewell and Cambridge Mathematics." *Historical Studies in the Physical Sciences* 11: 1–48.

———. 1980b. "Woodhouse, Babbage, Peacock and Modern Algebra." *Historia Mathematica* 7: 389–400.

———. 2004. "Peacock, George (1791–1858)." In *Oxford Dictionary of National Biography*. Oxford: Oxford University Press.

Benbaji, Yitzhak, and Menachem Fisch. 2004. "Through Thick and Thin: A New Defense of Cultural Relativism." *Southern Journal of Philosophy* 62: 1–24.

———. 2005. "Factuality without Realism: Normativity and the Davidsonian Approach to Meaning." *Southern Journal of Philosophy* 43: 505–30.

Ben Menahem, Yemima (ed.). 2005. *Hilary Putnam*. Cambridge: Cambridge University Press.

Berkson, William. 1995. "Two Commandments for the Humanities." In Jarvie and Laor 1995, 1:221–33.

Bjorken, James, and Sidney Drell. 1964. *Relativistic Quantum Mechanics*. New York: McGraw-Hill.

Bloom, Harold. 1973. *The Anxiety of Influence: A Theory of Poetry*. Oxford: Oxford University Press.

Bloor, David. 1981. "Hamilton and Peacock on the Essence of Algebra." In Mehrtens, Bos, and Schneider 1981, 202–32.

Bos, Henk J. M. 2001. *Redefining Geometrical Exactness: Descartes' Transformation of the Early Modern Concept of Construction*. New York: Springer-Verlag.

Boyer, Carl B. 1968. *A History of Mathematics*. New York: J. Wiley.

Brandom, Robert B. 1994. *Making It Explicit: Reasoning, Representing, and Discursive Commitment*. Cambridge, MA: Harvard University Press.

———. 2000a. *Articulating Reasons: An Introduction to Inferentialism*. Cambridge, MA: Harvard University Press.

——— (ed.). 2000b. *Rorty and His Critics*. Malden, MA: Blackwell.

———. 2002. *Tales of the Mighty Dead: Historical Essays in the Metaphysics of Intentionality*. Cambridge, MA: Harvard University Press.

———. 2009. *Reason in Philosophy: Animating Ideas*. Cambridge, MA: Harvard University Press.

Brown, Daniel. 2012. "William Rowan Hamilton and the Uses of Poetry for Science." *Studies in Romanticism* 51 (4): 475–501.

———. 2013. *The Poetry of Victorian Scientists: Style, Science and Nonsense*. Cambridge: Cambridge University Press.

Bunge, Mario (ed.). 1964. *The Critical Approach to Science and Philosophy*. New York: Free Press.

Buss, Sarah, and Lee Overton (eds.). 2002. *Contours of Agency: Essays on Themes from Harry Frankfurt*. Cambridge, MA: MIT Press.

Buttmann, Günther. 1970. *The Shadow of the Telescope: A Biography of John Herschel.* Trans. B. E. J. Pagel, ed. and introd. David S. Evans. New York: Charles Scribner's Sons.

Byrne, Alex. 1996. "Spin Control: Comment on John McDowell's *Mind and World.*" *Philosophical Issues* 7: 261–73.

Campbell-Kelly, Martin. 1994. Introduction to Babbage 1864, edited with a new introduction. New Brunswick, NJ: Rutgers University Press.

Cannon, Walter F. 1964. "William Whewell, F.R.S.—Contributions to Science and Learning." *Notes and Records of the Royal Society of London* 19: 176–91.

Carnap, Rudolph. 1937. *The Logical Syntax of Language.* London: Kegan Paul, Trench Trubner and Co.

Clarke, Stanley G., and Evan Simpson (eds.). 1989. *Anti-Theory in Ethics and Moral Conservatism.* Albany: SUNY Press.

Clock, Daniel A. 1964. "A New British Concept of Algebra, 1825–1850." Ph.D. diss., University of Wisconsin.

Close, Frank. 2010. *Antimatter.* Oxford: Oxford University Press.

Cohen, Gerald A. 2000. *If You're an Egalitarian, How Come You're So Rich?* Cambridge, MA: Harvard University Press.

Cohen, L. Jonathan. 1989. "Belief and Acceptance." *Mind* 98: 367–89.

———. 1991. "Should a Jury Say What It Believes or What It Accepts?" *Cardozo Law Review* 13: 465–83.

———. 1992. *An Essay on Belief and Acceptance.* Oxford: Clarendon Press.

Collingwood, Robin G. 1939. *An Autobiography.* Oxford: Oxford University Press.

Collins, Harry M. 1985. *Changing Order: Replication and Induction in Scientific Practice.* London: SAGE Publications.

Collins, Harry M., Robert Evans, and Michael Gorman. 2007. "Trading Zones and Interactional Expertise." *Studies in History and Philosophy of Science* 38 (4): 657–66.

Creath, Richard. 2010. "The Construction of Reason: Kant, Carnap, Kuhn, and Beyond." In Domski and Dickson 2010, 493–510.

Crowe, Michael J. 1994. *A History of Vector Analysis: The Evolution of the Idea of a Vectorial System.* New York: Dover.

Curtis, Ronald C. 1986. "Are Methodologies Theories of Scientific Rationality?" *British Journal for the Philosophy of Science* 35: 135–61.

Damerow, Peter, Gideon Freudenthal, Peter McLaughlin, and Jürgen Renn (eds.). 2004. *Exploring the Limits of Preclassical Mechanics. A Study of Conceptual Development in Early Modern Science: Free Fall and Compounded Motion in the Work of Descartes, Galileo, and Beeckman.* 2d ed., enlarged and revised. New York: Springer.

Davidson, Donald. 1982. "Paradoxes of Irrationality." In Wolheim and Hopkins 1982, 289–305.

Deligiorgi, Katerina (ed.). 2006. *Hegel: New Directions.* Chesham: McGill-Queen's University Press.

DiSalle, Robert. 2002. "Reconsidering Kant, Friedman, Logical Positivism, and the Exact Science." *Philosophy of Science* 69 (2), 191–211.

———. 2006. *Understanding Space-time: The Philosophical Development of Physics from Newton to Einstein.* Cambridge: Cambridge University Press.

———. 2010. "Synthesis, the Synthetic A Priori, and the Origins of Modern Space-Time Theory." In Domski and Dickson 2010, 523–51.

Domingues, João Caramalho. 2008. *Lacroix and the Calculus*. Basel: Birkhauser.

Domski, Mary. 2013. "Observation and Mathematics." In Anstey 2013, 144–68.

Domski, Mary, and Michael Dickson (eds.). 2010. *Discourse on a New Method: Reinvigorating the Marriage of History and Philosophy of Science*. Chicago: Open Court.

Dubbey, John M. 1978. *The Mathematical Work of Charles Babbage*. Cambridge: Cambridge University Press.

Ducheyne, Stephen. 2011. "Kant and Whewell on Bridging Principles between Metaphysics and Science." *Kant Studien* 102 (1): 22–45.

———. 2014. "Whewell's Philosophy of Science." In Mander 2014, 77–88.

Duhem, Pierre. 1954. *The Aim and Structure of Physical Theory.* Trans. Phillip Wiener. Princeton: Princeton University Press. Originally published as *La théorie physique, son objet et sa structure*. 2d ed. Paris: Chevalier et Rivière, 1914.

———. 1980. *The Evolution of Mechanics.* Trans. Michael Cole. Alphen aan den Rijn: Sijthooff and Noordhoff. Originally published as *L'évolution de la mécanique*. Paris: A. Joanin, 1903.

Durand-Richard, Marie-José. 1990. "Genèse de l'algèbre symbolique en Angleterre: Une influence possible de John Locke." *Revue d'histoire des sciences* 43: 129–80.

———. 2011. "Peacock's 'History of Arithmetic': An Attempt to Reconcile Empiricism to Universality." *Indian Journal for the History of Science*. Special issue on Peacock's "History of Arithmetic" and the Indian context. Also available online: http://halshs.archives-ouvertes.fr/halshs-00530818/fr/

Earman, John. 1993. "Carnap, Kuhn, and the Philosophy of Scientific Methodology." In Horwich 1993, 9–36.

Enros, Phillip C. 1979. "The Analytical Society: Mathematics at Cambridge University in the Early Nineteenth Century." Ph.D. diss., University of Toronto.

———. 1981. "Cambridge University and the Adoption of Analytics in Early Nineteenth-Century England." In Mehrtens, Bos, and Schneider 1981, 135–48.

———. 1983. "The Analytical Society." *Historia Mathematica* 10: 24–47.

Feigl, Herbert ,and Michael Scriven (eds.). 1956. *The Foundations of Science and the Concepts of Psychology and Psychoanalysis*. Minnesota Studies in the Philosophy of Science, 1. Minneapolis: University of Minnesota Press.

Fisch, Menachem. 1985. "Necessary and Contingent Truth in William Whewell's Antithetical Theory of Knowledge." *Studies in History and Philosophy of Science* 16: 275–314.

———. 1987. "Il collegamento dei fatti: La nozione non necessaria di verità empirica di William Whewell." In Simili 1987, 39–63.

———. 1988. "A Physicists' Philosopher: James Clerk Maxwell on Mathematical Physics." *Journal of Statistical Physics* 51: 309–19.

———. 1991a. *William Whewell, Philosopher of Science*. Oxford: Clarendon Press.

———. 1991b. "A Philosopher's Coming of Age—A Study in Erotetic Intellectual History." In Fisch and Schaffer 1991, 31–66.

———. 1991c. "Antithetical Knowledge." In Fisch and Schaffer 1991, 289–309.

———. 1992. "Ein Blick vorwärts in die Vergangenheit: Ein Fall für den historischen Nominalismus." *Deutsche Zeitschrift für Philosophie* 40: 1279–94.

———. 1994a. "Towards a Rational Theory of Progress." *Synthese* 99, 277–304.

———. 1994b. "Trouble-Shooting Creativity." *History and Philosophy of the Life Sciences* 16, 141–53.

———. 1994c. "The Emergency Which Has Arrived: The Problematic History of 19th Century British Algebra—A Programmatic Outline." *British Journal for the History of Science* 27, 247–76.

———. 1997. *Rational Rabbis: Science and Talmudic Culture*. Bloomington: Indiana University Press.

———. 1999. "The Making of Peacock's *Treatise on Algebra*: A Case of Creative Indecision." *Archive for History of Exact Sciences* 54, 137–79.

———. 2003. "A Modest Proposal: Toward a Religious Politics of Epistemic Humility." *Journal of Human Rights* 2 (1): 49–64.

———. 2008. "Taking the Linguistic Turn Seriously." *European Legacy* 13 (5): 605–22. Special issue, "The Languages of Science and the Humanities," ed. Oren Harman.

———. 2010. "Toward a History and Philosophy of Scientific Agency." *Monist* 93 (4): 518–44. Special issue, "Philosophical History of Science."

———. 2013a. "How and Why I Write History of Science." *Science in Context* 26 (4): 573–85.

———. 2013b. "Science, Religion and Rationality—A Neo-Hegelian Approach." *Toronto Journal of Theology* 29 (2): 319–36.

———. 2014. "Babbage's Two Lives." *British Journal for the History of Science* 47 (1): 95–118.

———. 2015. "Criticism from Within *versus* Criticism from Without—My Debt to Yemima." *Iyyun—Jerusalem Philosophical Quarterly* 64: 160–71. In Hebrew.

———. 2017. "Deciding by Argument *versus* Proving by Miracle: The Myth-History of Talmudic Judaism's Coming of Age." *Toronto Journal of Theology*, Supplement 1: 103–27.

Fisch, Menachem, and Yitzhak Benbaji. 2011. *The View from Within: Normativity and the Limits of Self Criticsm*. Notre Dame, IN: University of Notre Dame Press.

Fisch, Menachem, and Simon Schaffer (eds.). 1991. *William Whewell: A Composite Portrait*. Oxford: Clarendon Press.

Frankenberry, Nancy K. (ed.). 2002. *Radical Interpretation in Religion*. Cambridge: Cambridge University Press.

Frankfurt, Harry G. 1971. "Freedom of the Will and the Concept of a Person." *Journal of Philosophy* 68: 5–20. Reprinted in Frankfurt 1988, 1–25. Pagination according to reprint.

———. 1982. "The Importance of What We Care About." *Synthese* 53: 257–90 Reprinted in Frankfurt 1988, 80–94.

———. 1987. "Identification and Wholeheartedness." In Schoeman 1987, 27–45. Reprinted in Frankfurt 1988, 159–76. Pagination according to reprint.

———. 1988. *The Importance of What We Care About*. Cambridge: Cambridge University Press.

———. 1992. "On the Usefulness of Final Ends." *Iyyun—Jerusalem Philosophical Quarterly* 41: 3–19. Reprinted in Frankfurt 1999, 82–94. Pagination according to reprint.

———. 1999. *Necessity, Volition, and Love*. Cambridge: Cambridge University Press.

———. 2004. *The Reasons of Love*. Princeton, NJ: Princeton University Press.

———. 2006. *Taking Ourselves Seriously and Getting It Right*. Ed. Debra Satz, with comments by Christine M. Korsgaard, Michael E. Bratman, and Meir Dan-Cohen. Stanford: Stanford University Press.

Fraser, Craig G. 1989. "The Calculus as Algebraic Analysis: Some Observations on Mathematical Analysis in the 18th Century." *Archive for History of Exact Sciences* 39: 317–35.

———. 1990. "Lagrange's Analytical Mathematics, Its Cartesian Origins and Reception in Comte's Positive Philosophy." *Studies in the History and Philosophy of Science* 21: 243–56.

Freeman, Eugene (ed.). 1976. *The Abdication of Philosophy: Philosophy and Public Good—Essays in Honor of Paul Arthur Schilpp*. La Salle, IL: Open Court.

Friedman, Michael. 2001. *Dynamics of Reason: The 1999 Kant Lectures at Stanford University*. Stanford: CSLI Publications.

———. 2002a. "Kant, Kuhn, and the Rationality of Science." *Philosophy of Science* 69 (2): 171–90.

———. 2002b. "Geometry as a Branch of Physics: Background and Context for Einstein's 'Geometry and Experience.'" In Malament 2002, 193–229.

———. 2003. "Eckart Förster and Kant's *Opus Postumum*." *Inquiry* 46 (2): 215–27.

———. 2010a. "Synthetic History Reconsidered." In Domski and Dickson 2010, 571–813.

———. 2010b. "A Post-Kuhnian Approach to the History and Philosophy of Science." *Monist* 93 (4): 497–517.

Galison, Peter. 1997. *Image and Logic: A Material Culture of Microphysics*. Chicago: University of Chicago Press.

Garber, Daniel. 1986. "Learning from the Past: Reflections on the Role of History in the Philosophy of Science." *Synthese* 67: 91–114.

Gattei, Stephano. 2008. *Thomas Kuhn's "Linguistic Turn" and the Legacy of Logical Empiricism: Incommensurability, Rationality and the Search for Truth*. Burlington, VT: Ashgate.

Gaukroger, Stephen. 2001. *Francis Bacon and the Transformation of Early-Modern Philosophy*. Cambridge: Cambridge University Press.

Geertz, Clifford J. 1994. "The Strange Estrangement: Taylor and the Natural Sciences." In Tully 1994, 83–95.

Giunti, Marco. 1988. "Hattiangadi's Theory of Scientific Problems and the Structure of Standard Epistemologies." *British Journal for the Philosophy of Science* 39 (4): 421–39.

Gorman, Michael. 2002. "Levels of Expertise and Trading Zones." *Social Studies of Science* 32 (6): 933–38.

Grabiner, Judith V. 1997. "Was Newton's Calculus a Dead End? The Continental Influence of Maclaurin's *Treatise of Fluxions*." *American Mathematical Monthly* 104: 393–410.

———. 2002. "Maclaurin and Newton: The Newtonian Style and the Authority of Mathematics." In Withers and Wood 2002, 143–71.

Grattan-Guinness, Ivor. 1992. "Charles Babbage as an Algorithmic Thinker." *IEEE Annals of the History of Computing* 14 (3): 34–48.

———. 1996. "Numbers, Magnitudes, Ratios, and Proportions in Euclid's *Elements*: How Did He Handle Them?" *Historia Mathematica* 23 (4): 355–75.

Guicciardini, Niccolò. 1989. *The Development of the Newtonian Calculus in Britain 1700–1800*. Cambridge: Cambridge University Press.

———. 2003. "Three Traditions in the Calculus: Newton, Leibniz and Lagrange." In *Companion Encyclopedia of the History and Philosophy of the Mathematical Sciences*, 308–17. Baltimore: Johns Hopkins University Press.

———. 2004. "Dot-Age: Newton's Legacy in Eighteenth-century Mathematics." *Early Science and Medicine* 9 (3): 218–56.

Gunther, Robert T. 1937. *Early Science in Cambridge*. Oxford: Oxford University Press.

Habermas, Jürgen. 2000. "From Kant to Hegel: On Robert Brandom's Pragmatic Philosophy of Language." *European Journal of Philosophy* 8 (3): 322–55.

Hacking, Ian. 1979/1981. "Imre Lakatos's Philosophy of Science." *British Journal for the Philosophy of Science* 30 (4): 381–402. Extracts reprinted with revisions as "Lakatos's Philosophy of Science" in Hacking 1981, 128–43.

——— (ed.). 1981. *Scientific Revolutions*. Oxford Readings in Philosophy. Oxford: Oxford University Press.

Hankins, Thomas L. 1977. "Triplets and Triads: Sir William Rowan Hamilton on the Metaphysics of Mathematics." *Isis* 68 (2): 174–93.

———. 1980. *Sir William Rowan Hamilton*. Baltimore: Johns Hopkins University Press.

Harman, Oren, and Peter L. Galison. 2008. "Epistemic Virtues and Leibnitzian Dreams: On Science, Faith and the Humanities." *European Legacy* 13 (5): 551–75. Special issue, "The Languages of Science and the Humanities," ed. Oren Harman.

Hattiangadi, Jagdish N. 1978. "The Structure of Problems (Part I)." *Philosophy of the Social Sciences* 8: 345–65.

———. 1979. "The Structure of Problems (Part II)." *Philosophy of the Social Sciences* 9: 49–76.

Haugeland, John. 1998. *Having Thought: Essays in the Metaphysics of Mind*. Cambridge, MA: Harvard University Press.

———. 2005. "Reading Brandom Reading Heidegger." *European Journal of Philosophy* 13 (3): 421–28.

Herman, Barbara. 2002. "Bootstrapping." In Buss and Overton 2002, 253–74.

Horwich, Paul (ed.). 1993. *World Changes: Thomas Kuhn and the Nature of Science*. Cambridge, MA: MIT Press.

Hyman, Anthony. 1978. "Ancestral Figure of Computing." Review of Dubbey 1978. *Nature* 272: 779.

———. 1982. *Charles Babbage, Pioneer of the Computer*. Princeton, NJ: Princeton University Press.

———. 1989. *Science and Reform: Selected Works of Charles Babbage*. Cambridge: Cambridge University Press.

Jarvie, Ian C., and Nathanel Laor (eds.). 1995. *Critical Rationalism: The Social Sciences and the Humanities: Essays for Joseph Agassi*. 2 vols. Dordrecht: Reidel.

Kant, Immanuel. 1965 (1781). *Critique of Pure Reason*. Trans. Norman K. Smith. New York: Macmillan.

———. 1996 (1797). *The Metaphysics of Morals.* Trans. and ed. Mary Gregor with introduction by Roger J. Sullivan. Cambridge Texts in the History of Philosophy. Cambridge: Cambridge University Press.

———. 1997 (1788). *Critique of Practical Reason.* Trans. and ed. Mary Gregor with introduction by Christine M. Korsgaard. Cambridge Texts in the History of Philosophy. Cambridge: Cambridge University Press.

———. 1998 (1785). *Groundwork of the Metaphysics of Morals.* Trans. and ed. Mary Gregor with introduction by Christine M. Korsgaard. Cambridge Texts in the History of Philosophy. Cambridge: Cambridge University Press.

Koertge, Noretta. 2010. "How Should We Describe Scientific Change? Or: A Neo-Popperian Reads Friedman." In Domski and Dickson 2010, 511–22.

Koppelman, Ellaine. 1971/2. "The Calculus of Operations and the Rise of Abstract Algebra." *Archives for History of Exact Sciences* 8: 155–242.

Korsgaard, Christine M. 1996a. *The Sources of Normativity.* With G. A. Cohen, R. Geuss, T. Nagel, and B. Williams; ed. O. O'Neill. Cambridge: Cambridge University Press.

———. 1996b. *Creating the Kingdom of Ends.* Cambridge: Cambridge University Press.

———. 2006. "Morality and the Logic of Caring—A Comment on Harry Frankfurt." In Frankfurt 2006, 55–76.

———. 2009. *Self-Constitution: Agency, Identity, and Integrity.* Oxford: Oxford University Press.

Koyré, Alexandre. 1939. *Études galiléennes.* Paris: Hermann.

Kuhn, Thomas S. 1970. *The Structure of Scientific Revolutions.* 2d ed. International Encyclopedia of Unified Science, 2 (2). Chicago: University of Chicago Press.

———. 1976. "Mathematical versus Experimental Traditions in the Development of Physical Science." *Journal of Interdisciplinary History* 7: 1–31. Reprinted in Kuhn 1977, 31–65.

———. 1977. *The Essential Tension: Selected Studies in Scientific Tradition and Change.* Chicago: University of Chicago Press.

———. 1987. *Black-Body Theory and the Quantum Discontinuity, 1894–1912.* Chicago: University of Chicago Press. First published by Oxford University Press, 1978.

———. 1990. "The Road since 'Structure.'" In Kuhn 2000, 90–104.

———. 1993. Afterword. In Horwich 1993, 311–41. Reprinted in Kuhn 2000, chap. 11. Pagination according to reprint.

———. 2000. *The Road since "Structure": Philosophical Essays, 1970–1993, with an Autobiographical Interview.* Ed. James Conant and John Haugeland. Chicago: University of Chicago Press.

Lakatos, Imre. 1970. "Falsification and the Methodology of Scientific Research Programmes." In Lakatos and Musgrave 1970, 91–196. Reprinted in Lakatos 1978, 8–101.

———. 1971. "History of Science and Its Rational Reconstructions." In *Proceedings of the 1970 Biennial Meeting of the Philosophy of Science Association,* 91–136. Reprinted in Boston Studies in the Philosophy of Science, vol. 8, 91–136. Dordrecht: Reidel, 1971. Reprinted in Lakatos 1978, 102–38.

———. 1978. *The Methodology of Scientific Research Programmes.* Philosophical Papers, vol. 1 . Cambridge: Cambridge University Press.

Lakatos, Imre, and Alan Musgrave (eds.). 1970. *Criticism and the Growth of Knowledge.* Cambridge: Cambridge University Press.

Lambert, Kevin T. 2013. "A Natural History of Mathematics: George Peacock and the Making of English Algebra." *Isis* 104 (2): 278–302.

Lawrence, Christopher, and Steven Shapin (eds.). 1998. *Science Incarnate: Historical Embodiments of Natural Knowledge.* Chicago: University of Chicago Press.

Lipton, Peter. 2001. "Kant on Wheels." *London Review of Books* 19: 30–31.

Malament, David B. 2002. *Reading Natural Philosophy: Essays in the History and Philosophy of Science and Mathematics Presented in Honor of Howard Stein on the Occasion of His 70th Birthday.* Chicago: Open Court.

Mander, William J. (ed.). 2014. *Oxford Handbook of British Philosophy in the Nineteenth Century.* Oxford: Oxford University Press.

McDowell, John. 1978. "Are Moral Requirements Hypothetical Imperatives?" With Ian G. McFetridge. *Aristotelian Society* suppl. vol. 52: 13–29. Reprinted in McDowell 1998, 77–94.

———. 1979. "Virtue and Reason." *Monist* 62: 331–50. Reprinted in Clarke and Simpson 1989, 87–109. Reprinted in McDowell 1998, 50–75.

———. 1994. *Mind and World.* Cambridge, MA: Harvard University Press.

———. 1995. "Might There Be External Reasons?" In Altham and Harrison 1995, 387–98. Reprinted in McDowell 1998, 95–111. Pagination according to reprint.

———. 1998. *Mind, Value, and Reality.* Cambridge, MA: Harvard University Press.

———. 2006. "The Apperceptive I and the Empirical Self: Toward a Heterodox Reading of 'Lordship and Bondage' in Hegel's *Phenomenology.*" In Deligiorgi 2006, 33–48. Reprinted in McDowell 2009b, 147–65.

———. 2007. "On Pippin's Postscript." *European Journal of Philosophy* 15 (3): 396–410. Reprinted in McDowell 2009b, 185–203.

———. 2009a. *The Engaged Intellect: Philosophical Essays.* Cambridge, MA: Harvard University Press.

———. 2009b. *Having the World in View: Essays on Kant, Hegel, and Sellars.* Cambridge, MA: Harvard University Press.

McMackin-Garland, Martha. 1980. *Cambridge before Darwin: The Ideal of a Liberal Education, 1800–1860.* Cambridge: Cambridge University Press.

Mehrtens, Herbert, Henk Bos, and Ivo Schneider (eds.). 1981. *Social History of Nineteenth-Century Mathematics.* Boston: Birkhauser.

Mischel, Theodore (ed.). 1977. *The Self: Psychological and Philosophical Issues.* Lanham, MD: Rowman and Littlefield.

Mueller, Axel, and Arthur Fine. 2005. "Realism, beyond Miracles." In Ben Menahem 2005, 83–124.

Nagel, Thomas. 1996. "Universality and the Reflective Self." In Korsgaard 1996a, 200–209.

Nový, Luboš. 1973. *Origins of Modern Algebra.* Trans. Jaroslav Tauer. Prague: Academia.

Nye, Mary Jo, Joan L. Richards, and Roger H. Stuewer (eds.). 1992. *The Invention of Physical Science: Intersections of Mathematics, Theology and Natural Philosophy since the Seventeenth Century—Essays in Honor of Erwin N. Hiebert.* Dordrecht: Kluwer.

Phillips, Christopher. 2006. "Robert Woodhouse and the Evolution of Cambridge Mathematics." *History of Science* 44 (143): 69–93.

Pippin, Robert B. 2007. "McDowell's Germans: Response to "On Pippin's Postscript." *European Journal of Philosophy* 15 (3): 411–34. Reprinted in Pippin 2015, 63–90. Pagination according to reprint.

———. 2008. *Hegel's Practical Philosophy: Rational Agency as Ethical Life.* Cambridge: Cambridge University Press.

———. 2011. *Hegel on Self-Consciousness: Desire and Death in the "Phenomenology of Spirit."* Princeton, NJ: Princeton University Press.

———. 2015. *Interanimations: Receiving Modern German Philosophy.* Chicago: University of Chicago Press.

Popper, Karl R. 1963. *Conjectures and Refutations: The Growth of Scientific Knowledge.* London: Routledge and Kegan Paul.

———. 1976. "The Myth of the Framework." In Freeman 1976, 23–48.

———. 1994. *The Myth of the Framework: In Defence of Science and Rationality.* London: Routledge.

Putnam, Hilary. 1976. "'Two Dogmas' Revisited." In Ryle 1976, 202–13.

———. 1994. *Words and Life.* Cambridge, MA: Harvard University Press.

Pycior, Helena M. 1981. "George Peacock and the British Origins of Symbolical Algebra." *Historia Mathematica* 8: 23–45.

———. 1982. "Early Criticism of the Symbolical Approach to Algebra." *Historia Mathematica* 9: 392–412.

———. 1983. "Augustus De Morgan's Algebraic Work: The Three Stages." *Isis* 74: 211–26.

———. 1984. "Internalism, Externalism, and Beyond: Nineteenth-century British Algebra." *Historia Mathematica* 11: 424–41.

Quine, Willard V. O. 1951. "Two Dogmas of Empiricism." *Philosophical Review* 60: 20–43.

———. 1990. *In Pursuit of Truth.* Cambridge, MA: Harvard University Press.

Quinn, Philip L. 2004. Review of Harry G. Frankfurt's *The Reasons of Love.* Notre Dame Philosophical Reviews: http://ndpr.nd.edu/news/23713-the-reasons-of-love/

Rees, Graham. 1985. "Quantitative Reasoning in Francis Bacon's Natural Philosophy." *Nouvelle de la république des lettres* 1: 27–48.

———. 1986. "Mathematics in Francis Bacon's Natural Philosophy." *Revue internationale de philosophie* 159: 399–426.

Richards, Joan L. 1980. "The Art and Science of British Algebra: A Study in the Perception of Mathematical Truth." *Historia Mathematica* 7: 343–65.

———. 1987. "Augustus De Morgan, the History of Mathematics, and the Foundations of Algebra." *Isis* 78: 7–30.

———. 1991. "Rigor and Clarity: Foundations of Mathematics in France and England, 1800–1840." *Science in Context* 4 (2): 297–319.

———. 2007. "In Search of the 'Sea-Something': Reason and Transcendence in the Frend/De Morgan Family." *Science in Context* 20 (3): 509–36.

———. 2011a. "God, Truth and Mathematics in Nineteenth-century England." *Theology and Science* 9 (1): 53–74. First published in Nye, Richards, and Stuewer 1992, 51–78.

———. 2011b. "'This Compendious Language': Mathematics in the World of Augustus De Morgan." *Isis* 102: 506–10.

Richardson, Alan. 2002. "Narrating the History of Reason Itself: Kuhn, Friedman, and a Constitutive A Priori for the 21st Century." *Perspectives on Science* 10 (3): 253–74.

Rorty, Richard. 1979. *Philosophy and the Mirror of Nature*. Princeton, NJ: Princeton University Press.

———. 1989. *Contingency, Irony, and Solidarity*. Cambridge: Cambridge University Press.

Rothblatt, Sheldon. 1968. *The Revolution of the Dons: Cambridge and Society in Victorian England*. London: Faber and Faber. Reissued with new introduction, Cambridge: Cambridge University Press, 1981. Pagination according to reprint.

Rouse-Ball, Walter W. 1889. *A History of the Study of Mathematics at Cambridge*. Cambridge: Cambridge University Press.

Ryle, Gilbert (ed.). 1976. *Contemporary Aspects of Philosophy*. Stocksfield, UK: Oriel Press.

Schaffer, Simon. 1994. "Babbage's Intelligence: Calculating Engines and the Factory System." *Critical Inquiry* 21 (1): 203–27.

———. 1996. "Babbage's Dancer and the Impresarios of Mechanism." In Spufford and Uglow 1996, 53–80.

Schoeman, Ferdinand (ed.). 1987. *Responsibility, Character and Emotions: New Essays in Moral Psychology*. Cambridge: Cambridge University Press.

Schweber, Samuel S. 1981. *Aspects of the Life and Thought of Sir John Frederick Herschel*. New York: Arno Press.

Sellars, Wilfrid. 1956. "Empiricism and the Philosophy of Mind." In Feigl and Scriven 1956, 253–329. Reprinted with introduction by Richard Rorty and study guide by Robert Brandom. Cambridge, MA: Harvard University Press, 1997. Pagination according to reprint.

Shapin, Steven. 2008. *The Scientific Life: A Moral History of a Late Modern Vocation*. Chicago: University of Chicago Press.

Simili, Raffaella (ed.). 1987. *L'epistemologia di Cambridge (1850–1950)*. Bologna: Il Mulino.

Skinner, Quentin. 1974. "Some Problems in the Analysis of Political Thought and Action." *Political Theory* 2 (3): 277–303.

Smith, Crosbie W., and Norton M. Wise (eds.). 1989. *Energy and Empire: A Biographical Study of Lord Kelvin*. Cambridge: Cambridge University Press.

Smith, Jonathan, and Christopher Stray (eds.). 2001. *Teaching and Learning in Nineteenth-century Cambridge*. Woodbridge, UK: Boydell and Brewer.

Snyder, Laura J. 2006. *Reforming Philosophy: A Victorian Debate on Science and Society*. Chicago: University of Chicago Press.

———. 2011. *The Philosophical Breakfast Club: Four Remarkable Friends Who Transformed Science and Changed the World*. New York: Broadway Books.

Spufford, Francis, and Jennifer Uglow (eds.). 1996. *Cultural Babbage: Technology, Time and Invention*. London: Faber.

Sreenivasan, Gopal. 1998. "Interpretation and Reason." *Philosophy and Public Affairs* 27: 143–71.

Stout, Jeffrey. 2002. "Radical Interpretation and Pragmatism: Davidson, Rorty and Brandom on Truth." In Frankenberry 2002, 25–52.

———. 2004. *Democracy and Tradition*. Princeton, NJ: Princeton University Press.

Stray, Christopher. 2001. "A Parochial Anomaly: The Classical Tripos 1822–1900." In Smith and Stray 2001, 31–44.

Taylor, Charles. 1964. *The Explanation of Behaviour*. London: Routledge and Kegan Paul.

———. 1971. "Interpretation and the Sciences of Man." *Review of Metaphysics* 25 (1): 3–51. Reprinted in Taylor 1985b, 15–57. Pagination according to reprint.

———. 1977. "What Is Human Agency?" In Mischel 1977, 103–35. Reprinted in Taylor 1985a, 15–44. Pagination according to reprint.

———. 1981. "The Concept of a Person." In *Social Theory and Practice*. The B. N. Ganguli Memorial Lectures, Delhi: Oxford University Press. 1983, 48–67. Reprinted in Taylor 1985a, 97–114. Pagination according to reprint.

———. 1985a. *Human Agency and Language*. Philosophical Papers, vol. 1. Cambridge: Cambridge University Press.

———. 1985b. *Philosophy and the Human Sciences*. Philosophical Papers, vol. 2. Cambridge: Cambridge University Press.

———. 1989. *Sources of the Self: The Making of Modern Identity*. Cambridge, MA: Harvard University Press.

———. 2016. *The Language Animal: The Full Shape of the Human Linguistic Capacity*. Cambridge, MA: Harvard University Press.

Tsou, Jonathan Y. 2003. "A Role for Reason in Science." *Dialogue* 42: 573–98.

———. 2010. "Putnam's Account of Apriority and Scientific Change: Its Historical and Contemporary Interest." *Synthese* 176 (3): 429–45.

Tully, James (ed.). 1988. *Meaning and Context: Quentin Skinner and His Critics*. Cambridge: Polity Press.

Tully, James (ed.) 1994. *Philosophy in an Age of Pluralism: The Philosophy of Charles Taylor in Question*. Cambridge: Cambridge University Press.

Verburgt, Lukas M. 2015. "'A Terrible Piece of Bad Metaphysics?'—Towards a History of Abstraction in Nineteenth and Early Twentieth Century Probability Theory, Mathematics and Logic." Ph.D. diss., University of Amsterdam.

———. 2016. "Duncan F. Gregory, William Walton and the Development of British Algebra: 'Algebraical Geometry,' 'Geometrical Algebra' and Abstraction." *Annals of Science* 73 (1): 40–67.

———. Forthcoming. "Duncan Farquharson Gregory and Robert Leslie Ellis: Second-Generation Reformers of British Mathematics." *Intellectual History Review*.

Walzer, Michael. 1987. *Interpretation and Social Criticism*. Cambridge, MA: Harvard University Press.

———. 1988. *The Company of Critics: Social Criticism and Political Commitment in the Twentieth Century*. New York: Basic Books.

Warwick, Andrew. 1998. "Exercising the Student Body: Mathematics and Athleticism in Victorian Cambridge." In Lawrence and Shapin 1998, 288–326.

———. 2001. "The Analytical Revolution from Below: Private Teaching and Mathematical Reform in Georgian Cambridge." In Smith and Stray 2001, 5–30.

———. 2003. *Masters of Theory: Cambridge and the Rise of Mathematical Physics*. Chicago: University of Chicago Press.

Wilkes, Maurice V. 1990. "Herschel, Peacock, Babbage and the Development of the Cambridge Curriculum." *Notes and Records of the Royal Society of London* 44 (2): 205–19.

Withers, Charles W. J., and Paul Wood (eds.). 2002. *Science and Medicine in the Scottish Enlightenment*. East Linton, UK: Tuckwell Press.

Wittgenstein, Ludwig. 1958. *Philosophical Investigations*. 3d ed. New York: Macmillan.

———. 1974. *On Certainty*. Ed. G. E. M. Anscombe and G. H. von Wright; trans. Denis Paul and G. E. M. Anscombe. First published 1969; reprinted with corrections and indexes. Oxford: Blackwell.

Wolheim, Richard, and James Hopkins (eds.). 1982. *Philosophical Essays on Freud*. Cambridge: Cambridge University Press.

Yeo, Richard. 1993. *Defining Science: William Whewell, Natural Knowledge and Public Debate in Early Victorian Britain*. Cambridge: Cambridge University Press.

INDEX